Leks

MONOGRAPHS IN
BEHAVIOR AND ECOLOGY

Edited by John R. Krebs and
Tim Clutton-Brock

Leks

JACOB HÖGLUND AND
RAUNO V. ALATALO

Princeton University Press
Princeton, New Jersey

Library of Congress Cataloging-in-Publication Data

Höglund, Jacob, 1958–
Leks / Jacob Höglund and Rauno V. Alatalo.
p.   cm — (Monographs in behavior and ecology)
Includes bibliographical references (p.   ) and index.
ISBN 0-691-03728-0 (CL : alk. paper)
ISBN 0-691-03727-2 (PA : alk. paper)
1. Lek behavior. I. Alatalo, Rauno V.
II. Title. III. Series.
QL761.H64   1995
591.56′2—dc20   94-41218

Contents

List of Drawings at Part Openings

(all drawings by Dafila Scott)

Part I.   Fallow deer (*Cervus dama*) show extraordinary intraspecific variation in mating systems and some populations are known to lek (Langbein & Thirgood 1989). On leks, males show tremendous variation in mating success (Clutton-Brock et al. 1988).

Sage grouse (*Centrocercus urophasianus*) is an example of a bird species that leks throughout its range (Hjorth 1970). This species also exhibits great variation in male mating success (Gibson and Bradbury 1985).

Part II.   Great snipe (*Gallinago media*) fight to gain access to lek territories (Höglund and Robertson 1990b), thus male-male competition is a component of sexual selection in this species.

Black grouse (*Tetrao tetrix*) females sample males by repeated visits to their territories and finally mate with one (Alatalo et al., subm.). This is an example of the other major component of sexual selection, female choice.

Part III.   Dung flies (*Scatophaga stercoraria*) congregate on patches of fresh cow dung (Parker 1970). In this species all aggregation is explained by resources (dung), thus dung flies do not lek. However, this system is valuable as a starting point in thinking about the evolution of mating aggregations (leks).

Ruffs (*Philomachus pugnax*) are unique among birds in that males come in two morphs, probably genetically determined: satellites and residents (van Rhijn 1991). These morphs behave differently. Residents defend lek territories, whereas satellites engage in short-term associations within resident territories. Both can have access to females. Alternative mating tactics are important in understanding the evolution of leks.

Part IV.   Birds of paradise (genus: Paradisea) constitute one fo the most extreme examples of elaboration of male adornment. In birds of paradise, lek behavior correlates with diet: fruit-eating species tend to lek, whereas insect feeders have other kinds of mating systems (Beehler and Pruett-Jones 1983).

This drawing of a black grouse lek illustrates the two main unresolved questions about lekking. First, why do females show mate preferences for certain males? Second, why are males aggregated despite the intense competition for the access of females?

Preface

Two main aspects of lekking are of interest to evolutionary biologists. The first is that leks seem to offer ample oppurtunity for sexual selection. Many studies of sexual selection in lekking species have asked questions about adaptation and how selection molds phenotypes, and seek to answer how bizarre and superficially maladaptive structures, such as elaborate display behavior and enlarged secondary sexual morphological features, have evolved. The second aspect is that lek mating systems have evolved in some species and not in others. Why is this so, and why do these males aggregate in special areas? In this book we attempt to address both of these aspects of lekking.

The book is organized into four parts. In part I (chapters 1 and 2) we describe and define what a lek mating system is and review the taxonomic occurrence of leks. In part II (chapters 3–6) we look at sexual selection in lekking species. In part III (chapters 7–9) we focus on the evolution of leks. In part IV (chapter 10) we draw the main conclusions from the earlier parts and aim at a synthesis. We also point out new areas of research in the field.

We have not attempted to write a full history of the research on lekking or the development of the ideas that explain lekking. Rather, our aim here is to present the theory of lekking and the most relevant data to illustrate both aspects of it. We have tried to present facts as objectively as possible. However, when it comes to speculation and interpretation of facts, our own views on any specific matter will obviously be biased. While we emphasize the present state of the art, we do not disregard all the ideas, speculations, and field data that over the years have contributed to the development of this fascinating research field. We deeply respect the painstaking efforts of the biologists who conducted the pioneering studies on leks that have made it possible for us to write this book. Our own field experience comes from birds and amphibia, which causes some bias in the book. However, this bias is mostly due to the fact that most studies on lekking species have been carried out on birds. Birds are highly suitable for field observations, as are ungulates, which have recently also been studied in great detail. Other groups, such as insects, are particularly suitable for experimental manipulations, and given this possibility, we hope that more entomological research will be devoted to lekking and related activities. Leks in different groups of animals are very variable in their organization, and for this reason we think it is necessary to study all kinds of animals to fully understand leks.

It is particularly satisfying for researchers from Finland and Sweden to write a text on leks. It is a bit like leks being brought home again. Leks of capercaillie, ruff, great snipe, and in particular black grouse are relatively common in our countries, and observing leks has a long tradition there. Already in 1908, the zoologist L. A. Jägerskiöld published a surprisingly detailed and accurate book on sexual dimorphism, sexual displays, and leks. Furthermore, the word "lek" was probably brought to England from Sweden by the hunter and naturalist L. Lloyd (1867), who was born in England but lived a large part of his life in Sweden as a businessman in Gothenburg. Lek comes from the verb "leka," which means to play. Its original meaning was probably "to run, to jump." In early Scandinavian it meant "performing with rapid movements," probably because of the simultaneous display by males and the frequent fighting in the arenas. Locals talk about "leken," which is the lek itself, and "leka," which is the behavior of the animals on the lek. However, leks are not called "lekar" in all of Sweden; in some provinces "spel" is used, which means "game." Darwin had read Lloyd's book on Swedish and Norwegian game birds, and in *The Descent of Man* both "lek" and "spel" are mentioned (Darwin 1871: lek, e.g., p. 370; spel, p. 375).

Part of this text was written while JH was on sabbatical leave at the University of East Anglia, where Bill Sutherland was the kindest of hosts; we learned the game theory approach to analyzing animal distributions from him. Anders Berglund, Torbjörn Ebenhard, Robert Gibson, Susan Hannon, Matti Hovi, Cheryl Jones, Pekka T. Rintamäki, Birgitta Sillén-Tullberg, Bo G. Svensson, Bill Sutherland, Staffan Ulfstrand, and Fredrik Widemo either read all of or large parts of the drafts of the text and kindly commented on it and corrected some of the errors. Joni Aspi helped us to find many of the studies on insects. Staffan Andersson, Gwen Bachman, Andrew Balmford, James Deutsch, Peder Fiske, Robert Gibson, Thomas Hoffmeister, Hans Landel, Dov Lank, Rory Nefdt, Stephen Pruett-Jones, Bill Sutherland, and Dave Westcott provided us with unpublished material or pointed out references of which we were unaware.

For permission to reproduce material from published works, we are indebted to the authors (whose names appear in the captions to the relevant figures) and to the following publishers and journals: Academic Press (figs. 1.3b, 1.5, 2.2, 3.3a, 3.6, 3.25, 4.3, 4.6b, 6.3, 6.5, 8.2); *Evolution* (fig. 3.4); Oxford University Press (figs. 3.7, 5.3, 6.2, 7.5); E. J. Brill Publishers (fig. 6.6); Macmillan Magazines Ltd. (figs. 3.8, 3.9, 4.1a); American Ornithologists Union (figs. 1.3a, 3.27b, 3.27d); Blackwell Scientific Publications Ltd. (fig. 2.4); Springer-Verlag (figs. 1.2, 1.3d, 1.4, 2.3, 3.2, 3.3b, 3.5, 3.10,

3.27c, 6.1); the University of Chicago Press (figs. 2.1, 3.13, 8.3); the National Academy of Sciences (fig. 4.4); Cambridge University Press (figs 4.1b, 4.2); Paul Parey Verlag (fig. 1.3c).

JH was financially supported by the Swedish Natural Sciences Research Council (NFR), and RA by the Academy of Finland. NFR awarded us with a grant to cover the cost of the illustrations, which were drawn by Dafila Scott, and the figures, drawn by Astrid Ulfstrand.

PART I   Leks and Their Taxonomic Occurrence

1

What Are Leks?

1.1 Introduction

Some animals, such as the black grouse, mate in arenas, or so-called leks. The terms "lek" and "lekking behavior" were first used for the mating arenas of birds (L. Lloyd 1867), but more or less similar mating aggregations occur in a wide variety of taxa (chapter 2). The main attribute of leks is that the males are aggregated in one area and display close together. On a typical black grouse lek in Fenno-Scandia, 2–25 males display next to one another, and territory sizes of the central males are on the order of 10×10 m. The distances between the nearest neighboring groups of displaying males are about 2 km. The females visit the lek and are rather free to choose to mate with any male on the lek. After mating, females will leave the lek whereas the males will stay and continue to display toward other females. The female incubates the eggs and attends the young all by herself. Thus, males provide no resources, except the sperm necessary to fertilize the eggs, and no parental care. This pattern is typical of all organisms which lek, that is, which have a lek mating system. A striking feature of the black grouse mating system is the high variance in male reproductive success. Typically, in lek mating systems sexual selection is very strong. A few males get most of the matings while the majority get nothing at all.

The above paragraph shows that "lek" can mean three things. First, it refers to the place where males collect. Such sites are also called encounter sites, arenas, mating grounds, booming grounds, hooting grounds, cooing grounds, or dancing grounds (Hamerstrom and Hamerstrom 1958), but in most cases it is simply referred to as the lek. Second, to lek, lekking, or lekking behavior can refer to the display behavior of the males or the behavioral tendency of males to collect at a communal mating ground. Since sexual displays are not only found in animals that mate on leks, we will confine our discussion of lekking behavior to the behavioral tendency to join a mating aggregation. Third, "leks" can refer to the mating system of populations that mate on leks. Such mating systems are also called lek-promiscuity, lek-polygyny, or lek mating systems, and "lekkers" are those who lek.

A large part of this book is concerned with the question of why this mating system has evolved in some species or populations but not in others. We ask which selective pressures lie behind the evolution of lek mating systems, and which are the features of lek breeding species and populations. We will take note of the great diversity of lekking behavior in different species of animals when we try to understand why leks evolve. The rest of the book is devoted to the issue of sexual selection on leks. The theory of sexual selection has been under very intensive debate and development in recent years. Leks have been thought to provide one of the best systems for empirical tests of sexual selection models, since in lekking animals the resources that a male controls have been regarded as uninfluential for female mate choice. In non-lekking animals where males provide food or breeding sites for the female, female choice is primarily directed toward attaining the best resources, and there is no major problem in understanding how female choice operates. Conversely, in lekking animals the issue has been to explain female choice in the absence of any immediate effects on female fertility. In this book we will review the premises of this assumption and suggest areas in which we believe more research effort is needed.

In this introductory chapter we will define and describe what constitutes a lek. It is not immediately obvious how the lek should be defined, since lek-like mating aggregations are very variable in their characteristics. Therefore we use a rather loose definition that includes all kinds of male aggregations visited by females primarily for the purpose of mating. The rest of the chapter will serve as a description of the various types of leks and mating aggregations that resemble leks. We will also briefly compare leks with other mating systems, in particular with respect to what kind of sexual selection is predicted to be influential in each of them.

1.2 Definition of the Lek

Leks can loosely be defined as any aggregation of males that females visit only for the purpose of mating. However, animals can aggregate for a number of reasons (Bertram 1978), and we are here particularly interested in cases when animals are more aggregated than would be predicted by general resource distributions such as patchy habitats or localized food. Bradbury (1981) suggested four criteria that could be used to distinguish "classical" leks from other mating systems:

1. There is no male paternal care. Males contribute nothing to the next generation except gametes.
2. There is an arena or lek to which females come and on which most of the mating occurs. An arena is a site on which several males aggregate and does not fill the habitat normally used by the species for other activities such as feeding, roosting, etc.

3. The display sites of males contain no significant resources required by females except the males themselves. This stipulation includes food, water, roosts, nest sites, egg deposition sites, etc.

4. Females have an opportunity to select a mate when they visit the arena.

The first three criteria of no parental care, male aggregation in arenas, and lack of resources in male territories are generally believed to satisfy what most other researchers also mean by a lek (e.g., Oring 1982). The fourth criterion, of free female choice, has been more debated however (e.g., Beehler and Foster 1988). As we will see later in this text, there are doubts as to whether females can choose freely in some species as well as to what extent male-male competition is important in shaping the mating patterns. The fourth criterion was originally included to distinguish leks from mating swarms of mayflies and other insects where females are believed to have low or no ability to discriminate among males (Bradbury 1985). This view of insect mating aggregations has been challenged, as females of some aggregating insect species may be able to choose among males, contrary to earlier belief (e.g., Lloyd 1979, Kimsey 1980, Lederhouse 1982, Alcock and Smith 1987). Furthermore, some researchers working with vertebrates have suggested that leks can evolve without female choice (e.g., Clutton-Brock et al. 1992). Thus it appears as if females of some vertebrate species may have less choice than previously thought, whereas the opposite may be true for many insect species.

While it has been difficult to decide to what extent female choice must be present for the mating system to be called a lek, the same problem applies more or less to all four criteria of Bradbury (1981). Variation in the level of aggregation has led to the use of the term "exploded lek" (Emlen and Oring 1977) for cases when males are too loosely aggregated according to the classical view of leks. Likewise, the presence of resources close to the lek has led to the use of the term "resource-based lek" (Alexander 1975), but Bradbury (1985) pointed out that this criterion should not be used too strictly. Least ambiguous is no doubt the first criterion stating the absence of paternal care. However, it is not useful in distinguishing lekking from other mating systems since there are a huge number of animal species without paternal care that do not mate on leks.

Some species, such as lekking gamebirds, may be close to perfectly fulfilling the four criteria. However, many other animals are so reasonably close to satisfying the requirements that one cannot avoid discussing them in the context of leks. The lesson is that each criterion in a definition has to be considered as a continuum rather than as a strictly categorizing variable (see Bradbury 1985). If one would like to take a strict view, it might seem that it is useless to speak about leks at all. However, such a view would distort most of the definitions of any phenomenon in behavioral studies. We are quite convinced that we have not misled readers of this book as to the scope of its content by using the word "leks" in the book title. Therefore the term is practical, and its practicality will be enhanced by admitting the problems inherent in having a single clear-cut definition for lek.

The use of the traditional terms, however, may also hamper the development of our comprehension of animal behavior. For instance, the tradition of categorizing animal mating systems as monogamy, polygyny, polyandry, and promiscuity (Lack 1968, Selander 1972, Wittenberger 1979) does not really admit that there are several aspects of mating systems (such as territoriality, parental care, bonds between mates, and paternity) that should be considered separately. Furthermore, each species and even each population can show tremendous variation in its mating pattern.

In addition, one may get really lost when one considers the phenomenon of helping and communal breeding (see Brown 1987) or lekking in the traditional and simplified context of the above-mentioned four mating system categories. Within these four categories, leks have been placed under "arena promiscuity" by Wittenberger (1979), who defined promiscuity as absence of prolonged association between the sexes and multiple matings by members of at least one sex. However, the word "promiscuity" is often used only to indicate that both males and females mate several times with different individuals (e.g., Krebs and Davies 1993). Emlen and Oring (1977) in their ecological classification of mating systems did not have any promiscuous category, but referred to lekking as male dominance polygyny. In many lekking species the majority of females copulate with a single male only, suggesting polygyny, but there are also many species where females commonly mate with multiple males, suggesting sexual promiscuity by both sexes. Since females may commonly mate with multiple males even in species with monogamous pair bonds (Birkhead and Møller 1992, 1993), it will be more useful to categorize mating systems only in terms of spatiality, pair bonding, and parental care. Lek mating systems may thus be defined as aggregated display with no pair bonding or paternal care.

So when is it practical to speak about leks? In our view a lek is the following:

Aggregated male display that females attend primarily for the purpose of fertilization.

Or, to amplify, we have tried to follow an operational definition of leks as a mating system where males are more aggregated at the suitable display habitat than could be expected from random male settlement, and where females visit these aggregations for the purpose of mating, not because of food or breeding sites within the aggregation.

We have thus excluded the criterion stating that there should be no paternal care, since it is not necessary. Given that the other requirements are met, this criterion is unlikely to eliminate many mating aggregations from lekking. Bradbury (1981) mentioned some possible cases for gulls and storks that may exhibit mate choice in arenas (called "clubs"), even if males later do take part in the care of their own young. Such systems are exceptional, and they do not fulfill the other criteria of the lek since females do not choose males only for the purpose of copulation. Lack of paternal care is a common feature of lekking populations, but there are many species that lack paternal care or a bond between the sexes and do not mate on leks. The other criterion excluded by us is the necessity of free female choice within the arena. As discussed already, this criterion has been debated, and we think that it is enough that females show some level of choice simply by mating on an arena rather than mating indiscriminately. In species where females show no specific mate-seeking activity, mating aggregations may arise in sites where females emerge, feed, or breed. Insects provide many such examples (Thornhill and Alcock 1983), but such aggregations are not leks if they are not visited by the females primarily for copulation.

Our definition of leks results in groups of species scattered throughout the animal kingdom (see chapter 2). The evolution of such mating aggregations should, in our mind, be scrutinized at the same time. Using a more stringent definition would put too much of the focus on the arbitrary ways of defining the phenomenon rather than explaining the mechanisms that underlie the evolution of mating aggregations. For those who think we misuse the term lek for many mating aggregations that should not be called leks, we suggest that while reading this book they think of all mating aggregations as leks.

In the next pages we will briefly illustrate the variability among different lekking systems. We will also discuss a few examples of borderline cases—so-called exploded leks, male aggregation on landmarks, and resource-based leks—in which researchers have been uncertain about calling the mating aggregation a lek.

1.3 Level of Aggregation: Exploded Leks

In some species, even if males are clustered in a suitable display habitat, they can still have fairly large territories. Such lek mating systems have sometimes been called *exploded* leks (Emlen and Oring 1977) or *dispersed* leks (Gilliard 1969) as opposed to more clumped male display in *classical* leks. However, it is not always clear to what extent this distinction is a useful one. The differences between exploded and classical leks are continuous rather than categorical and should thus be treated with caution. Still,

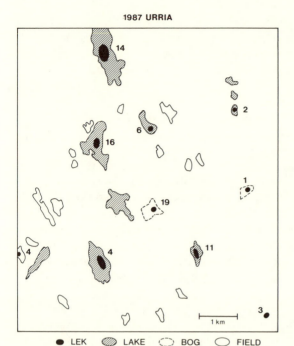

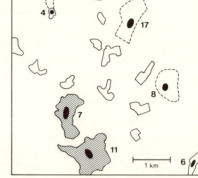

● LEK   ▨ LAKE   ⬭ BOG   ○ FIELD

Figure 1.1 The distribution of lekking sites of black grouse in two areas in our study site in central Finland. The numbers of males normally present during the main mating period are indicated. Apart from lakes, open bogs, and fields, lekking is also possible in forest clear-cuttings, as in the lower left corner of the map on the left.

undeniably, some lek species could be regarded as more dispersed than others. Ruffs and great snipe are both Palearctic lekking waders of about the same size (the mass of breeding males are about 210 g and 150 g, respectively). Yet the lekking territories of ruffs are only a few square meters (van Rhijn 1991), whereas those of the great snipe are about 100 square meters (Höglund et al. 1990a). However, both ruff and great snipe display on leks that have been considered to be classical leks (Hogan-Warburg 1966, Lemnell 1978).

In some cases it may be questionable whether males are so dispersed that it is not at all useful to speak about aggregations and leks. For instance, in a brood-parasitic finch, the village indigobird (Payne and Payne 1977), males sing on traditional call sites that are located a few hundred meters apart. Females visit several males in rapid succession and are courted by each one, but they eventually return to only one of the males to copulate. In such cases it may be difficult to ascertain whether males are really aggregated within the suitable display habitat. If they are, then we would regard them as lekking, otherwise the system is a case of solitary male display. One of the main focuses in this book will be to understand the mechanisms

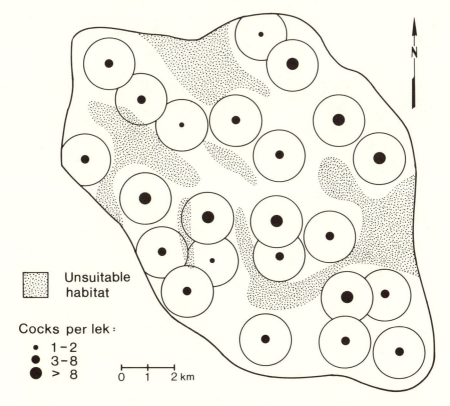

Figure 1.2 The spacing of capercaillie leks in Norway. Unsuitable habitat consists mainly of open peat bogs and lakes. A 1 km radius is drawn around each lek to indicate the space that might be occupied by territorial cocks of each lek. (From Wegge and Rolstad 1986)

that lead to the clumped male display (–lekking) and also to understand the causes of the variation in aggregation.

Another example of the variability in the level of male aggregations is provided by forest-inhabiting grouse. Black grouse is a species with classical leks that are well separated (fig. 1.1). In our study site in central Finland, most of the lekking takes place on ice-covered lakes, which is possible because matings usually occur before the ice melts. The other favored display habitat is open bogs, which are nowadays of limited number because of modern forestry management. Some of the leks are on fields and clear-cuttings that are available in large numbers. Open habitats are needed because in closed habitats goshawks have a very good chance of killing the males. Territory sizes in the central parts of larger leks are approximately 100 m^2. In contrast, the capercaillie has been regarded as having an exploded lek since average territory size is larger (400 m^2; Cramp and Simmons 1980). Quite opposite to the black grouse, it has been suggested that male capercaillie defend an area around the lek (fig. 1.2; Hjorth 1970,

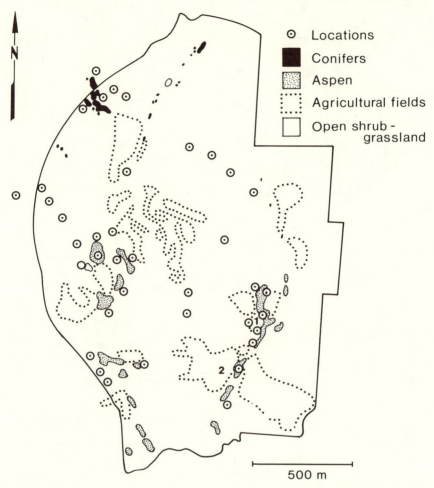

Figure1.3 Locations (dots) and vegetation types of areas where blue grouse males were usually hooting in an area in Washington state. Dark and light-shaded areas are conifers and aspen, respectively. Agricultural fields are outlined with dotted lines. (From Lewis 1985)

Wegge and Rolstad 1986). The display arena itself, however, consists of a relatively small area of the suitable habitat. In blue grouse, which inhabit North American forests, it has been suggested that males are aggregated to some degree (fig. 1.3; Lewis 1985). Even if distances between males are on the order of 100–200 m, some aggregation is apparent. However, the question is if there is aggregation within the habitat used by the species, and along the continuum from clumped to solitary display, is this species closer to solitary display? A clear example of solitary display is another American forest-living grouse, the spruce grouse, where males have no tendency at all to clump close together (Hjorth 1970).

1.4 Aggregation on Landmarks

One special type of male mating aggregation is found in species that use specialized habitat as encounter sites for mating. Earlier we mentioned that males should be aggregated within the suitable habitat before they can be considered to be lekking. However, one of the early explanations for lek-king was based on the idea that the suitable display habitat is limited and therefore males are aggregated (see Parker 1978a). The most obvious ex-amples come from the insects that aggregate on landmarks such as hilltops, forest clearings, sunspots, or treetops (see Thornhill and Alcock 1983, Al-cock and Smith 1987). For instance, two species of orchid bees, *Eulaema meriana* and *Euglossa imperialis*, have concentrations of male display ter-ritories in treefalls in the primary forests of Panama, but it is questionable if males are aggregated within the treefall (fig 1.4a; Kimsey 1980).

One particularly well studied landmark species is the tarantula hawk wasp (Alcock 1979a,b, 1981). Hawk wasps aggregate in the Sonoran desert of Arizona where males defend small trees and shrubs on hilltops. Females visit these aggregations to choose a mate and do not use the male territories further. Males do not assist the female once mating is over. These hawk wasps and many other hilltopping species may thus be viewed as parallels to leks in birds and other vertebrates. They fit well the broad definition of the lek we have chosen to apply, since females do visit male aggregations primarily for the purpose of mating. The special property of these leks is that certain features of the habitat such as the top of the hill are used to guide both males and females to the mating arenas.

Such landmark species have frequently been considered to be lekking, and we agree with this view insofar as specific habitat types are used to reduce the costs for females in searching for males. However, if there is no tendency for males to be aggregated at only some of these specific habitat types, the evolution of such aggregations is likely to be different from ag-gregations within the specific habitat type. Unfortunately, insect studies have not considered whether all the suitable habitat types are used by males or if only some of them have males. Also, in the bower building fish *Cyr-tocara eucinostomus*, males are aggregated but within a certain range of water depth (McKaye et al. 1990).

Many bowerbirds were initially considered to be lekking (Gilliard 1969, Vellenga 1970, Cooper and Forshaw 1979), but nowadays they are usually thought to have solitary display sites (e.g., Pruett-Jones and Pruett-Jones 1982). No doubt males are not particularly clumped, and at the very best they represent exploded leks if they are to be regarded as having leks at all. One well-studied species is MacGregor's bowerbird, which breeds in New Guinea. Pruett-Jones and Pruett-Jones (1982) showed that male territories, rather than being clumped, are regularly distributed on the ridges of the

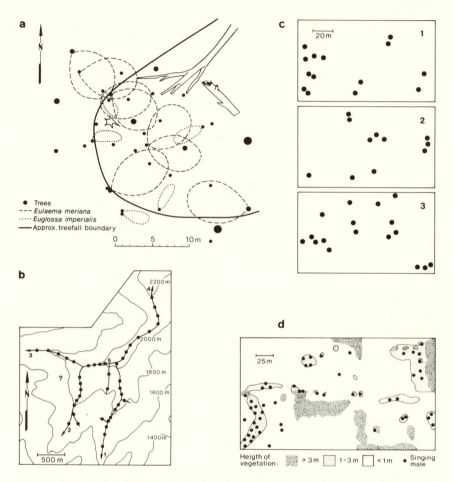

Figure 1.4 Spacing of displaying males in various species. (a) Orchid bees *Eulaema meriana* and *Euglossa imperialis* in a treefall in Panama (from Kimsey 1980). (b) MacGregor's bowerbird in New Guinea. Dots indicate bower sites, and bold lines the major and minor ridge lines (from Pruett-Jones and Pruett-Jones 1982). (c) Position of calling field cricket *Gryllus integer* males at different times of the year in the same field (from Cade 1981). (d) Positions of singing bushcricket *Tettigonia viridissima* males in relation to habitat type (from Arak and Eiriksson 1992).

montane forest (fig. 1.4b). This is somewhat parallel to insects that use landmarks; in fact, it resembles hilltopping since ridges may be used as conventional encounter sites for mating.

Another similar example comes from crickets, in which the term "lekking" has only seldomly been used. In fact, the emphasis has often been on the regularity of nearest-neighbor distances (Gwynne and Morris 1983). However, in many cases males are aggregated, but it is only within the

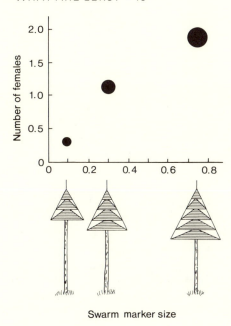

Figure 1.5 The mean number of *Empis borealis* females in sex-role-reversed aggregations swarming at experimental swarm markers (spruces) of different size. Areas of the dots are proportional to the number of males that visited per minute. (From Svensson and Petersson 1992)

aggregations that the nearest-neighbor distances are regular like the ones observed in the field cricket *Gryllus integer* (fig 1.4c; Cade 1981). The situation is similar in most of the lekking species. In fact, in most of the territorial lekking species an analysis of nearest-neighbor distances would probably show nonrandom overdispersion within each aggregation. However, in bushcrickets *Tettigonia viridissima*, males are quite saturated within one specific habitat type (fig 1.4d; Arak and Eiriksson 1992), and thus there is no true tendency to lek.

A nice example of the use of landmarks comes from the dance fly *Empis borealis* (see section 2.3 for this exceptional reversal of sex roles), where females are aggregated in aerial swarms close to small spruces (Svensson and Petersson 1992). Experiments where three different types of spruces were used indicated that female swarms were larger where the swarm marker was larger (fig. 1.5).

1.5 Resource Availability: Resource-based Leks

When males display close to or at resources used by females, for example at feeding sites, they are sometimes said to display at resource-based leks (Alexander 1975). In birds, this has been suggested to occur in some hummingbirds (see Stiles and Wolf 1979), a group of species that relies on nectar feeding. Groups of displaying males may be found near patches of flowers on which a particular species has specialized. Also, bellbirds and epauleted bats may court females from within territories that contain resources needed by females, but males do not use controlled access to these resources as leverage in obtaining females (Snow 1973, Bradbury 1981, 1985). Resource-based leks are, however, mainly confined to insects, at least it is in insects that the term has usually been used.

Thornhill and Alcock (1983) review a number of insect species that aggregate at feeding sites. In none of these cases is it clear if males are more clumped than would be predicted by resource distribution. We will later see that some models of lek evolution rely heavily on the distribution of resources. Males in classical lek species may use the spatial distribution of environmental resources used by females to select their lekking sites where the encounter rates with females presumably are maximized. Therefore, it is not meaningful beforehand to exclude any mating aggregations that are close to or at resources, such as food or nesting sites, from classical leks. As Bradbury (1985) stated, it is not the absence of resources within territories which should distinguish lek from non-lek species; it is the degree to which males regulate access to those resources to obtain matings that matters.

In some insects, such as the caddisfly *Mystacides asareus*, mating swarms are found near places where females are about to emerge (Petersson 1989). Similarly, insect aggregations are often found at sites where females oviposit their eggs (Thornhill and Alcock 1983). For instance, in the butterfly *Coenonympha pamphilus*, males are aggregated close to sites where females lay their eggs (fig. 1.6a; Wickman 1985, 1986). If such aggregations are to be considered as leks, however, it is necessary to show that males do not directly aggregate at the sites where most females emerge, where they are to oviposit their eggs, or where the females are feeding. Before such mating aggregations can be called leks, it should be verified that females, at least to some degree, do actively move to visit mating aggregations for the purpose of fertilization, and that such visits are not solely determined by the need to choose a certain site for ovipositing or feeding.

Note that the tendency of males to have mating swarms close to the sites where females occur does not contradict the lek definition. In fact, one of the most famous models of lek evolution, that is the hotspot model (Bradbury et al. 1986; section 7.2), is based on the idea that males aggregate in response to the density of females. For instance in the ruff, lekking sites are typically situated close to the ponds where females feed or in sites where females are likely to trespass while moving between feeding sites (fig. 1.6b; Höglund et al. 1993). Proximity to oviposition sites is particularly common in animal groups where fertilization is external, such as Amphibia and most fishes.

When resources used by females are outside male territories, there is no problem of treating the aggregations as leks. What makes resource-based leks of some hummingbirds and many insects different from a black grouse lek is that certain features of male territories, such as the number of flowers that a male controls, could be a critical resource for females. It is therefore possible that it is not the male but the resources that he controls that may be of concern when females choose with whom they mate. This consideration is particularly important in all aspects of lekking that relate to mate choice. In this case the difference between the lekking species and the species that have been considerd to have territorial polygyny becomes unclear. While hummingbird females do not nest in mating territories, they may be dependent on the food they can find there.

In some birds with a mating system based on territorial polygyny, such as the red-winged blackbird, females nest in male territories but may feed elsewhere. Recent DNA-fingerprinting analyses indicate that females may freely copulate with males other than the territory owner (Gibbs et al. 1990), and in many cases paternal care is missing or greatly reduced. No doubt, the mechanisms of sexual selection, female choice, and male clumping will in such cases approach the classical lek. As a case in point, we should recall that the most famous and influential experiment by Andersson (1982a) during the recent explosion of sexual selection studies was not made in a lekking species but in a species with territorial polygyny. Female long-tailed widowbirds nest in male territories, and they chose to nest in territories where the male's tail was elongated even if territorial characteristics were likely to be more influential on female breeding success and female choice. It is therefore practical to eliminate from lek systems all cases where essential amounts of food, nesting sites, or egg-laying sites are within male territories. In such systems these external resources will also have an effect on female choice and therefore they should be treated together with mating systems where male resources are essential.

A recent study by Wagner (1992) on razorbills illustrates the complexity of mating systems. This species is a colonially nesting seabird, traditionally regarded as monogamous with a true pair bond, and with both sexes taking care of the nest and the chicks. However, apart from normal intrapair copulations, both sexes attend mating arenas in the vicinity of the colony for extrapair copulations, and these aggregations function like a typical lek. Therefore, in this species the same individuals have both a monogamous pair bond and take part in lekking close to the nesting site.

While the availability of food or nest sites in male territories can be determined relatively easily, other resources that influence female survival and breeding success are difficult to ascertain. Safety from predation is a factor that influences female movements in basically the same way as the need to go to a place to feed, lay eggs, or nest. Risk of predation is one of the factors that has been suggested as a cause of male clumping (see

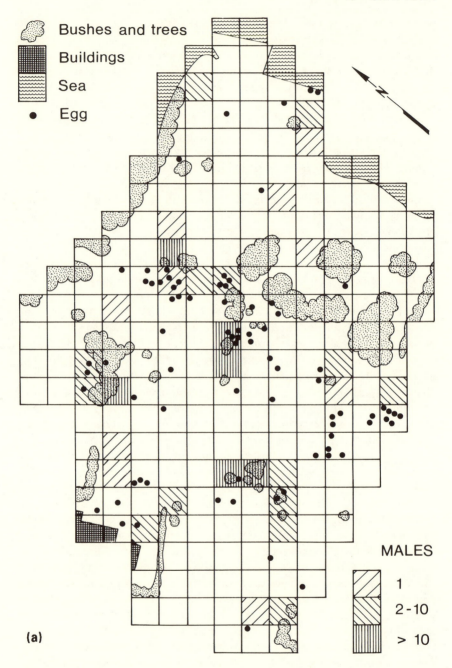

Bushes and trees

Buildings

Sea

● **Egg**

MALES

1

2-10

> 10

(a)

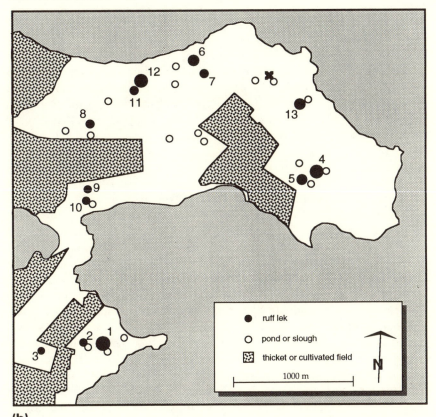

(b)

Figure 1.6 Location of lekking sites in relation to resources. (a) The distribution of station-
ary male butterflies (*Coenonympha pamphilus*) in relation to egg-laying sites of females
(from Wickman 1985, 1986). (b) The location of ruff leks in relation to ponds used as
feeding sites. The sizes of dots are proportional to the number of residents associated
with each lek (from Höglund et al. 1993).

Bradbury and Gibson 1983; section 7.2) and also as one of the factors that
may influence the choice of females on the lek. Traditionally this possibility has been taken as one possible feature of the lekking system. We agree
with this tradition for two reasons. First, as we will discuss later, predation
risk is not likely to be the major force driving male clumping and female
choice (section 7.3). Second, it would be extremely difficult to use a definition of a phenomenom that involves a mechanism that is so difficult to
estimate as predation risk.

The last argument applies even more clearly to the side effects of copulation on female breeding success and survival. Aside from sperm, diseases
and parasites may also be potentially transmitted during copulation. If this

is the case one could speak about disease avoidance as a resource. Likewise, the quality of sperm in fertilizing the eggs could be regarded as a resource and, finally, so could the genetic constitution of the sperm that influences offspring fitness. These are the important factors that are likely to drive female choice in lekking systems, and in this context we do not consider them as resources. Rather, it is the absence of the most obvious resources external to males, such as food and egg-laying sites, that are critical for the definition of the lek. In these cases, females visit males mainly for purposes other than fertilization. In our view, if females attend male aggregations for the purpose of fertilization, such aggregations are leks even if the sexual activities may involve immediate fertility effects.

Perhaps most problematic are the cases when males provide nuptial gifts during copulation. Contrary to much of the earlier tradition, we are prone to include such mating aggregations as leks in particular cases when the nuptial gifts are not essential for females. Studies in orthopterans have indicated that the quality of nuptial gifts may influence female fertility (Gwynne 1984, Butlin et al. 1987). However, recent work on bushcrickets indicate that nuptial gifts may not be as important as energy or nutrient sources as had been thought (Wedell and Arak 1989). It may be hard to see the logical difference between cases where females choose among clumped males to receive high-quality nuptial gifts together with sperm and cases when females try to avoid risk of disease transmission or try to achieve high-quality sperm for succesful fertilization. In all these cases, females visit the male aggregation for the purpose of fertilization, and they are not limited in their mate choice by availability of food or laying sites. However, if nuptial gifts are essential as sources of food for egg-laying females, the system should not be called lekking. On the other hand, if nuptial gifts serve as ornaments or if they have some other function not related to nutrient or energy acquisition by females, then these aggregations may be called leks.

1.6 Handling Time

One typical feature of leks is the brief female handling time per male. On a typical bird or mammal lek the insemination of any female is almost instant, and even the time it takes to convince a female to mate is not very long. Thus the minimum time between consecutive matings, H, is low (Sutherland 1985a,b, 1987). The consequence of this is that, apart from extreme occasions when many females visit a lek simultaneously, a mating male will not be drawn out of the breeding population and will not, by mating, miss many future breeding possibilities. Therefore the number of breeding males and, thus, potential fathers will be fairly constant over the season. This is important in terms of the extreme skew in male mating success that is typical of leks. In species where H is low, high variance in male mating success is expected as an effect of mere chance (Sutherland 1985a). In frogs, toads, newts, and many insects, males have to leave the breeding aggregation for extended periods of time in order to fertilize the female (e.g., Hedlund and Robertson 1989). The male handling time per female is thus much longer in these species, which restricts the mating options for subsequent females.

Breeding aggregations of anurans are usually classified as *explosive* or *prolonged* (Wells 1977, Arak 1982). In explosive breeders, breeding is very fast, often within a week, and male and female attendance at the breeding aggregations is highly synchronized. Explosive breeding aggregations are characterized by intense scramble competition for access to mates, there is usually no mate choice or territoriality, and the intensity of sexual selection depends entirely on the sex ratio at spawning and the length of the breeding season (Höglund 1988).

In prolonged breeders, on the other hand, breeding usually takes place during several weeks, female arrival at the breeding ponds is highly unsynchronized, and thus the operational sex ratio (Emlen and Oring 1977) on any given night is biased toward males. Prolonged breeders may defend territories in which females lay their eggs (e.g., bullfrogs; Howard 1983), and thus territory quality is probably more important than male characteristics in mate choice. Such systems therefore do not resemble leks. In other prolonged breeders, such as the natterjack toad, males defend so-called *acoustic territories*, a space around themselves in where they will not tolerate other conspecific males (Arak 1983b). Males are not found on the same place on subsequent nights, but "carry" their territories with them. However, the fertilization of any given female usually withdraws the male from the breeding population for an entire night, and spawning can in rare cases proceed for more than one night (J. Höglund, pers. obs.). Anuran choruses and breeding assemblages where the females do not breed in territories controlled by males are treated in this book as leks.

1.7 Stability of Arenas and Territories

In many cases leks are often found at traditional sites. We know of leks that have been in the same place for decades, as, for example, in the great snipe (Höglund and Robertson 1990a). Another feature of some leks, especially of vertebrate species, is that males often hold permanent territories. The same male is often found on the same part of the lek day after day and sometimes even between seasons. This aspect of time-fixed territoriality is sometimes not found in other vertebrates such as frogs and newts (Arak

1982, Hedlund and Robertson 1989) and in insect swarms (e.g., Petersson 1989). Territoriality was used by Baker (1983) to distinguish insect mating aggregations into swarms, which lack territoriality, and leks, in which territoriality occurs. In this book we do not use this criterion to distinguish leks from other mating systems.

There will always be some stability in the mating arenas of lekking species, since they are sites where, first, males aggregate and, second, females visit for mating. In the case of insects that swarm in sites where the sunlight reaches through the canopy, this stability may be relatively short in time. In insects that use landmarks such as hilltops, the lekking sites will remain stable without any true historical tradition, while in long-lived animals traditionality will pass over decades. However, the term "lek" cannot be used for aggregations of males and females that are not fixed to some particular site during mating and when females are already initially grouped with males for purposes other than mating. An example of such aggregations is the wildebeest that live in large mobile herds on the East African savanna. Matings take place within these herds during the normal mobile life of this antelope (Estes 1968). Likewise, many pelagic fish live continuously in mobile mixed-sex schools, and even if they are aggregated during mating they cannot be considered to be lekking.

1.8 Sexual Selection and Mating System

Much of the recent interest in lekking is derived from the general interest in sexual selection over the last years. The main unsolved problems of sexual selection deal with the importance of indirect selection on female choice, that is, selection that operates through differential success of the offspring without any effect on female fertility. The other modes of sexual selection—male-male competition (intrasexual selection) and the part of female choice (intersexual selection) that influences the reproductive success of the female (direct selection)—have been less debated. Since females in lekking systems appear only to achieve genes from their mates, it seems as if leks provide an ideal system to study indirect selection. In theoretical population genetics the feasibility of female choice of good genes has been debated (see section 4.2). That females of lekking species do seem to have mate preferences thus initially presented a paradox (the lek paradox of Taylor and Williams 1982).

While it is frequently emphasized that leks provide an excellent system for studying indirect sexual selection, we should recall that mating systems are not categorizable with respect to the sexual selection processes. Instead, the relative importance of each process varies between mating systems, and compared with other matings systems, indirect selection is likely to be relatively important in lekking animals.

Male-male competition is by definition particularly strong in female defense polygyny, a mating system where males directly control a harem of females. It is important also in territorial systems where males compete for high-quality territories to attract several females as envisaged in the polygyny threshold model (Orians 1969, Wittenberger 1979). Also, in lekking species males frequently fight among each other and, as we will see later, attaining dominance may have considerable impact on attaining copulations. In some lekking animals only the dominant male in the arena may copulate, and thus female choice is not always free. If males are displaying solitarily, male-male interactions are less likely but by no means absent.

Female choice may be present in all mating systems except perhaps in female defense polygyny and polyandry. In polyandry, the choice by males could, in contrast, be under selection. Direct benefits for female choice are particularly obvious in resource defense polygyny and monogamy, where females should be choosing males on the basis of resources in territories or on the quality of paternal care by the male. In the absence of such effects, the indirect effects, either through increased attractivness of sons or increased viability of the offspring of favored males, will be more important in the cases without a pair bond: in leks and in solitary display. However, direct benefits may be present in these species as well (Avery 1984, Reynolds and Gross 1990, Kirkpatrick and Ryan 1991), for example, in terms of the effects of sperm quality on fertility or in any effects that the mating decision may have on female condition or survival. In particular, the possibility of disease transmisson during copulation and the risk of receiving ectoparasites from the male are potentially important.

This brief overview of sexual selection mechanisms clearly illustrates the impossibility of studying only one of these mechanisms alone by choosing a representative mating system. The choice of lekking species for studies of female choice with indirect fitness benefits is practical because of the potential relative unimportance of direct selection or male-male competition. However, species with solitary display would be equally suitable, and from the viewpoint of sexual selection mechanisms they are very close to lekking systems.

1.9 Summary

Definitions should not restrict and hamper our analyses; they should instead help us to make clear what we are actually studying. The use of the term "lek" started with studies of birds, although male mating aggregations that females attend for the purpose of mating are a phenomenon also found in many other animal groups. The most popular definition of the lek has been that of Bradbury (1981), which has four criteria: (1) there is no male paternal care, (2) males are aggregated in an arena to which females come to mate, (3) the display site contains no significant resources for females, and (4) females have an opportunity to select their mate. However, defining the lek for general use in all animal taxa is not easy, since the criteria used for definition represent continuous variables. This is particularly true for the level to which females are free to choose between males within the male aggregation. Hence, we need a practical and less stringent definition of the lek. In this view, the lek is any case of *clumped display of males that females attend primarily for the purpose of mating*. The basic requirements are that males are more aggregated than could be expected from random settlement, and that fertilization is the primary reason for female visits to the aggregation.

There is great variation among species in the level of aggregation, and if males are dispersed, leks have been called *exploded* as opposed to *classical* leks. A special type of lek, common in insects, is found in species that use a landmark, such as hilltops, treetops, sunspots, or forest clearings as the conventional encounter sites where males aggregate to wait for visiting females. There is no clear-cut association between mating systems and the various sexual selection mechanisms. Female choice for indirect benefits through offspring, either in attractivnes of sons or offspring viability, is likely to be relatively important in lekking animals. However, female choice may also be based upon direct effects on female fertility, while male-male competition has varying levels of importance in different lekking species.

2

A Taxonomic Overview

2.1 Introduction

In this chapter we review the taxonomic occurrence of leks. What is a lekking species? Mating systems can be seen as the result of behavior of individuals in a population. As we will stress in chapter 9, different populations of the same species can have different mating systems. Moreover, the same population of a given species can change mating systems when conditions change. Finally, different individuals within a population can show differences in mating tactics, conditional on a number of factors such as size, age, and phenotypic quality. Therefore the mating system of the population will be a compromise of the behavior of the individuals that comprise it. In this chapter we consider a species to be lek breeding when a lek mating system occurs in at least one breeding population.

Leks occur in a large number of taxa. Table 2.A at the back of the chapter shows that it is evident that leks have evolved independently a number of times since there are many nonlekking species in all the phylogenetic categories listed. In birds, leks have evolved in at least fourteen independent cases (Höglund 1989). We are thus dealing with obvious multiple cases of convergent evolution.

Many more bird or ungulate species have been under behavioral observations than is the case for insects. Furthermore, the term "lek" was initially used for birds, while similar male aggregations in insects have been termed swarms, choruses, hilltoppers, or aggregations on landmarks. In spite of this we can be quite confident in saying that lekking, though widespread in different taxa, is not a particularly common mating system.

In table 2.A we have listed all species in which a lek mating system has been reported to date. This list is not complete, since as more species are being studied, more will be reported to lek. The incompleteness is also affected by intraspecific variation in mating systems (e.g., Langbein and Thirgood 1989; chapter 9). Many species have both lek mating and non-lek mating populations. Therefore, even in the more well studied species, lek mating may have been overlooked because the populations being studied have not mated on leks even though they may do so in other parts of the species range. In a few cases the available information is very brief and uncertain, and we have noted such cases in the table. Unfortunately, many published papers do not have enough information for us to be able to judge whether the system described is really a lek. Therefore, we must emphasize that our list should not be taken too literally and certainly not used for formal comparative analyses other than perhaps for birds and ungulates. With respect to birds and ungulates, the list is quite reliable for the species that have been studied. In some cases we suspect that a species is lekking but the authors have not used the term lek; we will mention such possibilities in the text but have not included them in the table.

While it is clear that there are species in which lek mating is conditional on a number of factors such as population density, habitat fragmentation, resource distribution, and other factors, it is less well established if obligate lek mating species exist. Even in extremely well studied species claimed always to lek, such as the black grouse, we know that lek mating is rare in some areas of the grouse's range. There are few published accounts of this because observations of non-lek mating in black grouse are seldom recorded, probably because solitary display is often performed in high trees and is therefore hard to see. Lek mating is indeed very common in black grouse, and seems less dependent on environmental factors than in the ungulates, for example (see chapters 7 and 9). It appears that there is a continuum from species that almost always lek to those that seldom lek.

If non-lek mating is the ancestral state and leks always evolved from non-lek mating systems, conditional lek mating may be the first stage. If this is true, then perhaps obligate lekking follows as an end point in species that have used a lek mating system for a long evolutionary time. Though suggestive, this idea is hard to test. Whatever the cause of intraspecific variation, the point here is that it is difficult to classify a species as lekking or not. In chapter 8 we will return to the reasons why one should expect intraspecific variation. The study of such species will provide important information about the reasons for the evolution of leks.

2.2 Basic Prerequisites for Lekking

Before we go further in our discussion of lekking in different groups, we will briefly look at the basic physiological and ecological circumstances that seem to be necessary for lekking to evolve. We will later return in more detail to these and other factors that explain lekking, but to be able to understand the occurrence of lekking in different taxa we must first know the basic prerequisites of lekking, which include these factors:

1. Males provide no parental care.
2. Males cannot monopolize resources to gain matings with females.
3. Internal fertilization takes place, allowing females to visit the leks for fertilization without ovipositing the eggs there. (Lekking is possible in

some cases of external fertilization if there is no association between the male chosen and the site of oviposition or if the female carries the eggs with her, as in some mouth-brooding cichlid fish.)

4. Mobility makes it possible for females to accept the cost of searching for mating aggregations and for males to avoid any excessive increase in predation risk.

5. Females must show some discriminatory behavior while searching for mates, or else females would mate at random and not in aggregations.

2.3 Arthropods Other Than Insects

There are no published accounts of lekking in arthropods other than insects. However, the mating aggregations of some fiddler crabs (*Uca*) could be similar to leks (Greenspan 1980, Zucker 1981, 1983, Christy 1982). However, we will not include them here as lekkers, since males defend breeding burrows, where, for example, female *Uca pugilator* incubate the eggs several days after mating (Christy 1978). A particularly well studied case of an arthropod breeding aggregation is the horseshoe crab *Limulus polyphemus*, which appears to be more similar to the breeding aggregations of explosive-breeding amphibians than to leks (Brockman 1990). A potential group that might have the tendency to lek are some wolf spiders that may aggregate at specific microhabitats for mating. For instance, male *Hygrolycosa rubrofasciata* aggregate on dry, sunny places within alder and birch marshes to acoustically attract females (Köhler and Tembrock 1987). Even if paternal care or prolonged pair bonding is typically absent in arthropods, the lack of mobility seems to exclude the possibility for leks.

2.4 Insects

In insects, swarming is a common and widespread mating behavior (e.g., Downes 1969, Sullivan 1981, Brittain 1982). Mating aggregations of flying males appear to have striking similarities to the leks of birds and mammals and choruses of frogs and toads (Bradbury 1985). Individuals of the nondiscriminating sex (in most cases the males) are aggregated, this sex provides no parental care, and in territorial species the territories contain no resources that could influence the mate choice of the visiting sex.

Whether or not the choosy sex can select its mates, lekking insects seem to differ from vertebrates. Bradbury (1985) concluded that in most insects "a female is immediately chased and mounted by one or more males and has no overt choice of a mate. Any non-randomness in male success is strictly due to intrasexual interactions between males." Bradbury points to several exceptions. In several species of *Drosophila*, males and females behave in a way very similar to lekking vertebrates (table 2.A). In the black swallowtail butterfly *Papilio polyxenes*, females to a large extent seem to be able to control with which males they mate (Lederhouse 1982). The same is true for Euglossine bees (Kimsey 1980) and lekking fireflies (Lloyd 1979).

Whether or not swarming insects should be included in a list of lekking species could be debated. As stressed by Bradbury (1985), it is not useful to argue whether a given taxon has leks or not. In the insect literature the most common practice is to call mating aggregations swarms and not leks. The term "lek" has usually been used for species where males display on vegetation having fixed territories or display sites. In table 2.A we listed only species that have explicitly been called lekking species. However, we are inclined to include swarming as lekking even if there is no territoriality and mate choice by females within swarms is frequently limited (Bradbury 1985, Petersson 1989). Therefore, the list of lekking insects should include many more of the aerial swarming species in Trichoptera, Ephemeroptera, Diptera, Hemiptera, Neuroptera, and Hymenoptera (Downes 1969, Sullivan 1981). However, swarms of many insects are often found near food resources, oviposition sites, or sites where virgin females emerge (Otte 1974, Alexander 1975). Thus in many cases it is unclear whether all the spatial clustering of the swarming males is determined solely by the ecological resource or habitat distribution, or if the male clustering to some extent is also determined by the males actively aggregating.

From table 2.A it can be seen that the majority of insects that have been termed to have leks belong to the order Diptera. Other insect orders in which lekking is reported in more than a few species are the butterflies and (Lepidoptera) and bees and wasps (Hymenoptera). Within the Diptera, most lekking species belong to the *Drosophila* species group and its allies. However, the behavior and mating systems of the vast majority of insects are still unknown, and any inferences on the significance of the taxonomic occurrence of lekking within the insects would be premature. In Hawaiian and Australian *Drosophila*, the leks have been explained as a way to avoid predation risks of displaying at the feeding sites (Parsons 1977a, Spieth 1978). However, recently it has been questioned whether the male aggregations on Australian bracket fungi are really leks, since female *Drosophila mycetophaga* may use fungi as a breeding resource (Hoffman and Blows 1992). In frugivorous tephritids (Diptera: Tephritidae) it has been suggested that in monophagous species, males would be expected to establish territories at the oviposition site, whereas in polyphagous species males would be unable to control access to oviposition sites (Shelly and Kaneshiro 1991).

Orthoptera are an insect order that has been quite extensively studied and where the word "lekking" has not been used even if aggregation of males is common (Gwynne and Morris 1983). A closer examination of some of the studies on this order reveals three problems. First, it is doubtful whether the spatial distribution of males and the observed patterns of female mate preferences for calling males can lead us to infer whether a species is lekking or not. Second, it is uncertain if the temporal distribution of the observed male calling activity corresponds to lekking. Third, it is uncertain to which degree nuptial gifts are essential as resources for females.

A general finding in orthopteran studies is that much of the spatial distribution of calling males can be explained in terms of specific habitat requirements and a heterogeneous environment (Greenfield and Shaw 1983). Habitat requirements could depend on factors such as the distribution of food plants (Greenfield et al. 1987) or sites optimal for male signaling (see fig. 1.4d; Arak and Eiriksson 1992). However, within patches where the habitat requirements are met, there appear to be mechanisms which both enhance further aggregation (i.e., female preference for aggregated males), and act in the opposite direction (i.e., male agonism that deters males from one another). Some evidence suggest that females of some orthopterans prefer to mate with aggregated males (Morris et al. 1978, Doolan and MacNally 1981, Shelly and Greenfield 1985). However, other studies fail to find this pattern even though males in the wild seem aggregated (see fig. 1.4c; Cade 1981). In the bushcricket *Tettigonia viridissima*, when males were experimentally clustered they suffered reduced mating success as compared to males who where spaced at distances found in nature (Arak et al. 1990). The latter result suggests that males maintain a regular acoustic spacing to minimize detrimental effects of agonistic behavior (see also Shaw et al. 1981). The male spacing observed in nature thus appears to be a compromise between female choice for aggregated males (if it occurs) and male agonism maintaining distances between males (Bailey and Thiele 1983).

The habit of orthopterans to call in unison and seemingly synchronize their behavior over time led Walker (1983) to suggest that this behavior was similar to "temporal lekking" and that it should be called a "spree." This is not similar to the "temporal lek" suggested by Campanella and Wolf (1974). In the latter study, a temporal lek is the spatial clustering of several males in a given territory. The first male to arrive on any given sequence of matings seems to be dominant over other males, but arrival, and subsequent dominance, changes between males and sequences. In this book spatially clustered males are called leks, and temporal clustering has been omitted.

In general, many of the prerequisites for lekking are fulfilled in insects (table 2.1), since there is no paternal care, fertilization is internal, and many insects are quite mobile. Indeed, lekking or a lekking type of behavior is not uncommmon in insects, and is particularly true for aerial swarmers. However, female control of mating may be relatively limited. Male aggregation close to resources is very common, but many times it is unclear

Table 2.1

An Overview of the Basic Prerequisites for Lekking to Evolve in Different Taxa

	Insects	Fish	Amphibia	Reptiles	Birds	Mammals
Male emancipation	++	+	++	++	+/–	++
No male resource control	+/–	+/–	+/–	+/–	+/–	+/–
Fertilization independence	++	–	– –	++	++	++
High mobility	+/–	+	–	– –	++	–
Female control of mating	–	–	–	+	++	+

NOTES: Negative signs indicate unfavorable conditions and postive signs favorable conditions. +/– indicates high variability in the condition within the taxa. Fertilization independence refers to internal as opposed to external fertilization or any other way that mate choice and oviposition site can be kept separate. Female control of mating refers to absence of forced copulations and the presence of active female choice behavior.

whether male aggregations should be called resource-based leks, or if they represent aggregations at emergence, ovipositing, or feeding sites. Also, male aggregation on landmarks is rather common, and hilltopping is thought to be connected with rarity (Thornhill and Alcock 1983). In rare species with dispersed feeding or oviposition sites, the use of landmarks will reduce the costs for females in finding the males.

A special case of a lek type of mating aggregation is the swarming of dance flies (Diptera, Empididae), which has been studied in detail in *Empis borealis* (Svensson and Petersson 1992). It is the females that gather to swarm while males carry nuptial gifts to them. Basically this mating system functions like a lek, and the evolution of the aggregations seems to be explicable by the same hypotheses as many of the typical male leks. However, it is questionable if males visit females primarily for fertilization. This is because males tend to choose larger females that are more fecund, and thus the system is closely comparable to the male aggregations where females receive nuptial gifts. If resources given by the visited sex are essential, we choose not to call such aggregations leks. However, such systems are very close to those cases in which females would choose between aggregated males for the purpose of receiving high-quality sperm. In any case, because of their special features the female aggregations of dance-flies might best be called "female leks" (see also Owens et al. 1994 for a possible parallel case in a bird, the Eurasian dotterel).

2.5 Fish

A large number of teleost fish species spawn communally (reviewed by Loiselle and Barlow 1978). Are these breeding aggregations similar to leks? The most common pattern of parental care in fishes is paternal care (Baylis 1981). Thus males in many fish species do provide females with resources essential for their immediate reproductive success. Even if males only defend a spawning site, the physical properties of this site may be important in terms of survival and developmental conditions for the spawn. In some fish species there is no parental care at all, and thus there is a possibility for lekking to evolve. However, in external fertilization, females may choose the site of spawning at the same time as they choose the male. Thus, with respect to mate choice on leks, most communally spawning fish do not resemble leks.

Furthermore, in species that shed the spawn into the open sea, such as the reef-dwelling blue-headed wrasses (Warner 1987), the position of territories on the reef is probably of some importance with respect to prevailing currents, and thus there is the probability that the fertilized eggs are swept away from predators on the reef out into the open sea. However, the impact of the spawning-site selection may not be essential, since experimental removals of males and females have indicated that traditionality influences the choice of display and spawning sites (Warner 1988, 1990). New sets of males will establish display sites other than those used previously by the removed set of males. This system would thus have a possibility for leks to evolve; in fact, some of the so-called initial-phase males, which are drab, act as sneakers, and parasitize so-called terminal-phase males in group spawnings (Warner 1990, Warner and Schulz 1992). However, the terminal-phase males in breeding coloration display where they are well separated. One possible factor limiting lek evolution is the ease with which any close-by subdominant males can sneak fertilizations during spawning. Therefore, it is particularly important for dominant males to keep other males away to avoid extensive "spatial spillover" as envisioned in the hot-shot hypothesis (chapter 7).

In the family Cichlidae, mouth brooding of the eggs and fry is common, and in some cichlid fish only females brood. Thus these species, in exception to most other fishes, fulfill the basic prerequisites for the evolution of lekking since females carry the fertilized eggs away from the site of fertilization, and such species also tend to lek (Lowe-McConnell 1959, Keenleyside 1991). These leks appear analogous to leks in birds and mammals. However, since communal spawning is common also in groups that have parental care, it is not clear if leks in cichlids evolved from such breeding aggregations and the maternal care is a later adaptation, or if leks evolved only in species where maternal care evolved. It seems as if in fish, lekking similar to what is found in birds and mammals has evolved only in species in which males have been liberated from parental care, particularly in female mouth brooders.

A particularly well studied lekking female mouth brooder is *Cyrtocara eucinostomus*. The males build bowers of sand, and females prefer males with tall bowers (McKaye 1983, McKaye et al. 1990). Males aggregate, and mean distances between bowers are about 2 m. Aggregation is at a depth of 6–7.5 m, probably because in deeper water there is the possibility

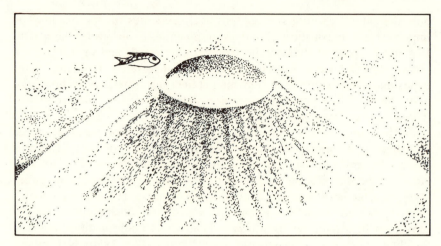

Figure 2.1 Bower of *Cyrtocara eucinostomus*, a female mouth breeder whose eggs are fertilized in the top crater of the sand mound. The length of the fish is 10 cm. (From McKaye et al. 1990)

of catfish predation. In this preferred depth, the habitat is saturated by bowers. Interestingly, bowers seem to have dual functions for males: the bowers are also designed so that eggs roll to the center of a crater at the top of the bower for fertilization (fig. 2.1). Thus it is less likely that sneak males will move in or that fishes on the substrate will be able to come in from the side and eat the eggs (Mc Kaye 1983).

2.6 Amphibians

In amphibians, lekking is most likely to be found among prolonged breeders (Wells 1977). In such species females arrive at the breeding areas asynchronously over a prolonged time period (from several weeks to months). In species where males are aggregated, the mating system often resembles leks. Frog and toad choruses are seldom referred to as leks in the herpetological literature, however, and therefore the listing of lekking anurans is scanty and incomplete. Another reason for this difficulty is intraspecific— and even intrapopulational—variance in the propensity of lekking. This problem is revealed by an example from the North American bullfrog.

In bullfrogs males defend small territories within ponds during the breeding season. In these territories males call and females visit calling males in order to mate. In 1965 and 1966, a pond in Michigan, USA, where circa two hundred male bullfrogs were breeding, was studied by Emlen (1968, 1976). Emlen concluded that the mating system exhibited by this population was similar to a lek: "Mating requisites of importance to females (such as oviposition sites) [do not] constitute resources that can be controlled or defended economically by males. Competition thus takes the form of direct interactions between males, leading to the spacing out of individuals within the chorus in accordance with physical and/or social dominance. The result is a communal display ground or lek" (Emlen 1976).

A few years later, in 1975–1978, another researcher studied the same species in the same pond (Howard 1978a,b, 1983, 1984.). During Howard's study, the number of males in the pond was much lower (range of 26–38 in different years). The main difference with respect to the mating system between Emlen's and Howard's studies was that Emlen found mated males and females leaving the chorus to oviposit the eggs, whereas Howard found that females oviposited their eggs within male territories. Thus, in Howard's study, the females could have decided whether or not to spawn with a particular male on the basis of physical features of the territories, such as shelter from wave action, predators, and microtemperature. Howard consequently did not classify the mating system as a lek but as territorial polygyny. If drastic changes in the mating systems of populations are commonplace, it becomes complicated to classify species as lekking.

Lek breeding has been reported in a number of other anurans. Among the best-known examples is the red-groined toadlet (Robertsson 1986a,b), where females choose among terrestrially calling males and carry their mates to a nearby pond where their eggs are fertilized. Among the widespread and diverse Bufonids, lek breeding has been claimed or can be inferred in Woodhouse's toad (fig. 2.2; Sullivan 1982), the raucous toad (Cherry 1993), and the natterjack toad (Arak 1988a). However, this list is most certainly incomplete, as the mating system of many Bufonids is still unknown. A very well studied case of a species that has been claimed to lek is the Pelobatid frog, the túngara frog. However, in this species males provide females with a foam nest in which they lay their eggs; hence female choice could be based on characters that are correlated with the ability of building good nests (Ryan 1985).

In treefrogs (Hylidae) lekking might occur in at least twelve species with prolonged breeding periods (see references in Whitney and Krebs 1975, Wells 1977, Fellers 1979), but to our knowledge only one Hylid has ever been specifically claimed to be lek-breeding: the gray treefrog (Sullivan and Hinshaw 1992). While species of this group call from positions removed from oviposition sites, it is typical that noncalling satellite males sit next to callers and attempt to intercept females (e.g., *Hyla cinerea*; Perrill et al. 1978). Noncalling satellites are common also in several species of toads (Arak 1983a). Aggregations with callers and satellites may provide the clearest example of the so-called hotshot mechanism (chapter 7), where

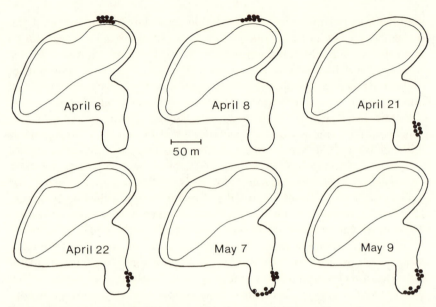

Figure 2.2 The clumped distribution of displaying *Bufo woodhousei* males in a pond on six different days of the prolonged mating season. (From Sullivan 1983)

less attractive males aggregate close to the attractive males to gain access to copulations.

As a rule it seems as if lekking is rare among anurans. Species with short breeding periods have mating systems characterized by scramble competition and thus do not resemble leks. In prolonged breeders, males often defend resources that are used by mating females. Such systems are best described as cases of territorial polygyny. In a few cases males do not defend resources, and here the mating system may be a lek. However, external fertilization easily leads to close associations between mate choice and the site of oviposition. In some species males display beside the oviposition sites, and these aggregations may be similar to resource-based leks. In many instances males do not defend the same space on consecutive nights (e.g., natterjack toads; Arak 1988a), and thus the territorial organization appears different from what is found in bird and ungulate leks. The success of satellite strategy in many species suggests that females cannot always control their mate choice well enough for classical leks to evolve.

2.7 Salamanders and Newts

In the Urodela (salamanders and newts), some populations of the crested newt have been reported to mate on leks (Hedlund and Robertson 1989), but to our knowledge no other species has been reported as lekking. The mating system of the most well studied congener of the crested newt, the smooth newt, seems similar when compared with the crested newt. Furthermore, there is no reason to believe that the mating systems of other *Triturus* species differ dramatically from the crested newt's. It thus appears as if lek mating systems may be more common than previously thought in this family.

During the breeding season, male crested newts migrate from their terrestrial habitats to the breeding ponds where they attain an aquatic life stage. Males can be found in small aggregations within the breeding pond, and these aggregations usually shift position from night to night. Females visit these aggregations and select their mates. When a male is selected, the pair usually moves away from the male aggregation, and courtship proceeds. As in the natterjack toad, fertilization takes considerable time and may last up to several hours. This system thus resembles vertebrate leks.

Are anurans and newts actually aggregated or only superficially so because they are forced to breed in ponds? Clearly, if males space themselves out while in the ponds, the breeding aggregation and the clustering is set by the size of the ponds. This would be different from a black grouse lek, where a group of males displaying in a bog would always be aggregated in only a small part of it. However, the male groups in Woodhouse's toad and crested newt are indeed more clumped than what would be set by the limits of the ponds in which they breed in (fig. 2.2). It is therefore important in each instance to carefully examine the distribution of males and what is setting the limits to their distribution.

Among salamanders, we have been unable to find records of lek mating systems. However, lek mating within this group may be more common than what has been reported.

2.8 Reptiles

In reptiles there are no reports of lekking species, with the possible exception of some anolid lizards (Stamps 1977, Fleishman 1992). Again, it is possible that lekking has been overlooked rather than being absent from the group. While internal fertilization should increase the potential for the evolution of lekking, the species within this group move slowly and are therefore likely to run great risks of predation on traditional aggregation sites. Furthermore, the costs for females to visit aggregations might be too high.

Marine iguanas mate in colonies, and the mating system is similar to that of seals (Trillmich and Trillmich 1984, Rauch 1985). Females rest in the territories of males and prefer territories with the best possibilities for thermoregulation both against the heat at noon and the chill in the morning and afternoon. Females preferentially copulate on the same territories where they rest. However, recent studies of marine iguanas suggest that females move to clusters of males to avoid sexual harassment. Marine iguanas may therefore constitute an example of where leks have evolved without female choice (sections 7.3 and 9.2).

2.9 Birds

Lek mating is best known in birds for a number of factors. One is that studies of mating systems have had a long tradition in ornithology, even though the mating systems of many bird species are still unknown. Also, the terminology and definitions of leks are borrowed from the ornithological literature. Lekking in birds was first well described by Nilsson (1824) and subsequent studies of leks were also on birds (e.g., Gadamer 1857, Lloyd 1867, Selous 1906, 1907). Thus, when describing a previously unknown mating system of a bird species, one would probably classify a doubtful case as lekking, whereas this would not be true in entomology, ichthyology, herpetology, or mammalogy.

In birds, lekking has been described in at least ninety-seven species (excluding the bowerbirds) in fourteen different families (table 2.A). Leks occur in such diverse genera as forest grouse (Tetraonidae) and hummingbirds (Trochlidae). Five families contain species that comprise 78% of all lek-mating birds. These are the forest grouse (8 species), the hummingbirds (Trochlidae, 14 species), cotingas (Cotingidae), and manakins (Pipridae) and birds of paradise (Paradisaeidae)(18 species, respectively). Within Tetraonidae, Cotingidae, Pipridae, and Paradisaedae, a large proportion of all species have been reported to mate on leks (table 2.2).

The occurrence of lekking is thus concentrated in a few avian families, suggesting a strong phylogenetic bias of the origin of leks in birds (see also Prum 1990, 1992). Furthermore, the number of lekking species could be contrasted against the estimate of the number of times lekking has evolved independently (97 vs. 14; Höglund 1989b; see also chapter 6).

There are two possible but not mutually exclusive explanations for this phylogenetic bias. First, in lek mating systems males are subjected to strong sexual selection. It has been suggested that because such selection may promote the evolution of different male traits in different subpopulations of the same species, rapid speciation could occur in lekking taxa (Lande 1980, 1987). Second, and probably more important, there may be strong phylogenetic biases on characters that make the evolution of leks possible. An obvious example of the latter is that maternal care of the young is a prerequisite for lekking. Accordingly, lekking can only evolve in groups were males do not assist the female in rearing the offspring, and related species are likely to share the same ecological circumstances that allow male emancipation and lek evolution (Prum, in press).

Table 2.2

Number and Proportion of Species Lekking in
Different Avian Families/Subfamilies

Family/Subfamily	No. of Lek Species	Proportion of Species Lekking
Tetraoninae	8	0.47
Phasianinae	2	0.01
Meleagrinae	1	0.50
Otidae	1	0.04
Scolopacidae	3	0.03
Psittacidae	1	0.01
Trochlidae	14	0.04
Indicatoridae	3	0.19
Cotingidae	18	0.21
Pipridae*	18	0.32
Tyrannidae	3	0.01
Oxyruncidae	1	1.00
Pycnonotidae	1	0.01
Paradisaedidae	20	0.48
Ploceidae	6	0.04

*Recent re-analysis of the taxonomy and phylogeny of Pipridae reveal
that all species of this family lek (Prum 1990, 1992, in press). However, the definition of leks used in these analyses differs from ours.

Male emancipation is an important prerequisite for lek evolution but cannot be the sole explanation since there are many species where males do not provide either parental care or resources and yet do not lek (McKitrick 1992). What causes male emancipation? In tropical lekking species, reduced clutch size due to high nest predation has been suggested as one explanation (Snow 1962, Snow and Snow 1979, Lill 1974, 1976, Beehler 1987). If reduced clutch size is selected for, the male is not needed in successfully raising the offspring. However, as pointed out by Beehler (1987), there is a cause-and-effect problem at least in birds of paradise. In this family, species in which both parents provide care tend to lay a clutch of two eggs, whereas in species in which only the female provides care the clutches consist of one egg. However, it is unclear whether reduced clutches selected for male emancipation or vice versa. In birds in general there seems to be a propensity toward reduced clutches in tropical species, suggesting that reduced clutch size came first (Ricklefs 1969). However, temperate lekking species may lay very large clutches, as exemplified by black grouse and other forest grouse in which clutch size can be up to nine or ten eggs. Even if the grouse are precocial and thus not comparable to altricial passerines, reduced clutch size alone as an explanation for male emancipation does not seem to be a general explanation. It seems that male emancipation is possible only if paternal care is not essential for the breeding success of females.

Other explanations for male emancipation relate to the diet of the offspring and to what extent the parents are needed to assist the young until independence. Male emancipation may thus be expected to be found in particular among precocial species, but contrary to this expectation many precocial species show biparental care (e.g., geese, ptarmigans, hazel grouse), while male emancipation is also commonly found in altricial species (birds of paradise, hummingbirds, bowerbirds). Male emancipation may not only be a consequence of how much care the young need but also to what extent male emancipation is limited by female diet (Beehler 1987). This argument predicts that female-only parental care has evolved in species in which the foraging niche is such that females can sustain the period when parental care is given without male assistance. Again, such ideas are hard to test by interspecific comparisons since the cause and effect is hard to reveal. Which sex will provide parental care is a problem of wide evolutionary significance and can be analyzed with the use of game theory. Such games have been coined "battle of the sexes" (Parker 1978b, Hammerstein and Parker 1987). In stable solutions of the battle-of-the-sexes games, either or both sexes can provide parental care, but it is generally thought that females become the care-giving sex because of anisogamy (Parker 1978b). In any case, it seems that among altricial birds male emancipation and lekking almost only occur in frugivores (birds of paradise, manakins, cotingas) or in nectar-feeding hummingbirds. It may be that in these cases the breeding success of females is not limited by food for the nestlings, as is typically true in insectivorous birds.

Given male emancipation, the distribution of resources is generally thought to determine which mating system becomes prevalent. This has been termed the *environmental potential for polygyny* (EPP)(Emlen and Oring 1977, Oring 1982). When resources essential for females are economically defendable, male resource defense polygyny is thought to evolve. When female groups can be followed and defended against other males, the prevailing mating system will be female defense polygyny. When resources are not defendable, the mating system will be based on male dominance (Oring 1982). Leks fall under male dominance polygyny according to this view and can be thought of as a default strategy when neither resources nor females can be monopolized (Oring 1982; see also Beehler 1987). The underlying reasons for the occurrence of leks are thus resource distribution, which in its turn will affect the distribution of females, and the EPP. Leks, as opposed to other mating systems based on male dominance, occur when female home ranges are large and overlapping, which requires moderate to high population densities (Oring 1982; see also chapters 7 and 8).

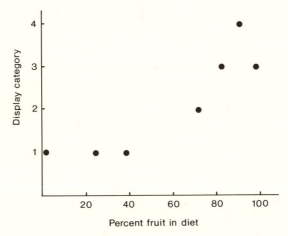

Figure 2.3 Display category in relation to percent fruit in the diet of seven species of birds of paradise. Display categories are: 4, clustered leks; 3, exploded leks; 2, solitary displays; and 1, defense of territories with resources. (After Beehler and Pruett-Jones 1983)

"Economically defendable resources" is a loose term that can almost be turned into a truism. It is very hard to define "economically defendable" first and then study the effects on the mating system afterwards. However, there have been some attempts to study this question in relation to diet.

In a comparison of display dispersion and diet in nine species of birds of paradise, there was a relationship between the extent of frugivory and the degree of clustering of displaying males (fig. 2.3; Beehler and Pruett-Jones 1983). The reason for such a relationship is that fruit resources are spatially and temporally clumped to such a degree that they are not defendable. It appears that resource defense occurs only in species whose diets are comprised of less than 50% fruit. However, the extent of frugivory cannot fully explain all levels of male dispersion. The shift from widely spaced non-resource-defending males to males tightly clustered on leks does not seem to be entirely explained by diet. Other considerations such as general habitat limitations and the relative significance of predation are possible additional factors of importance (Beehler and Pruett-Jones 1983, Beehler 1987).

A possible parallel to the birds of paradise may be found among the snipe. In the Scandinavian mountains the common snipe and the great snipe are found sympatrically. The great snipe is a lekking species whereas the mating system of the common snipe is not well studied, though, most evidence suggests some kind of resource defense polygyny (Tuck 1972). The great snipe seems to be almost entirely specialized to feed on lumbricids. The biomass of lumbricids on chalk-rich soils is ten times higher than on other soils (Løfaldli et al. 1992). The great snipe is almost exclusively

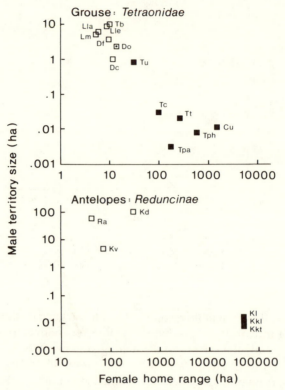

Figure 2.4 Male territory size in relation to female home-range size in (*top*) grouse (sub-family: Tetraoninae) and (*bottom*) antelopes (subfamily: Reduncinae). Filled squares are lekking species, open squares are non-lekking, and the dotted symbol indicates an ambig-uous case (see text). Tb = *Tetrastes bonasia*, Lla = *Lagopus lagopus*, Llc = *L. leucotus*, Lm = *L. mutus*, Df = *Dendragapus falcipennis*, Do = *D. obscurus*, Dc = *D. canadensis*, Tu = *Tetrao urogallus*, Tt = *T. tetrix*, Tc = *Tympanuchus cupido*, Tpa = *T. pallidicinctus*, Tph = *T. phasianellus*, Cu = *Centrocercus urophasianus*, Ra = *Redunca arundinum*, Kd = *Kobus defassa*, Kv = *K. verdoni*, Kl = *K. lechwe kaufawensis*, Kkl = *K. kob leucotis*, and Kkt = *K. kob thomasi*. (Adapted from Davies 1991; data from Bradbury et al. 1986 and Clutton-Brock 1989)

confined to chalk-rich soils in Scandinavia, where this otherwise rare bird can be found locally in high densities (J. A. Kålås, P. Fiske, and J. Höglund, unpubl. data). The common snipe, on the other hand, is a food generalist and is found in moderate densities in a wide array of habitats (Tuck 1972). It is thus possible that the differences in male dispersion are ultimately related to differences in feeding ecology of the two species. The reason for the great snipe's display on leks could be linked to the impor-tance of nondefendable food and the spacing and locally high densities of females in the habitat patches to which the species is confined.

The importance of female home-range size can be seen in figure 2.4 (top). When female home ranges are large, males have small and tightly clustered territories, that is, they are lekking. When females have large overlapping home ranges they can easily visit clumped male display arenas. What then is the underlying reason for the large size of female home ranges? All four species of grouse that live in open habitats (sage grouse, sharp-tailed grouse, lesser and greater prairie chickens) have very large female home ranges and are lekking species. Among the forest-living species black grouse and capercaillie have the largest female home ranges and are the only truly lekking forest grouse. Feeding habits may be important in determining the large home ranges, as well as the probable need of most of these species to flock outside the breeding time to minimize predation risk. Hazel grouse and ptarmigans are pair-bonding species with small home ranges. They are poor fliers who might suffer increased predation risk if they were to visit a lek, particularly since the hazel grouse relies on cryptic behavior to avoid predation.

In general, the possibilities for lekking are quite good in birds if there is male emancipation. Their high mobility, internal fertilization, and the relatively good control of matings by the females contribute to this tendency. However, male emancipation is not that common in the altricial species, but it is fairly common for males to control access to females through monopolization of high-quality territories, leading to resource defense polygyny. Somewhat surprisingly, many precocial species are monogamous even if it is doubtful if males can contribute much to female breeding success. In the case of ducks (Anatinae), females typically take care of the brood, yet monogamy is the prevailing mating system. One possible explanation is that since male ducks have penislike structures that make forced copulations possible, females need a mate to avoid continuous harassment from other males.

2.10 Ungulates and Other Mammals

In mammals, lekking is rare and is found mainly among ungulates, particularly in the African Bovidae, but also in the Cervidae, an African bat, an Australian marsupial, and in at least one population of dugongs (table 2.A).

In ungulates the habit of lekking is almost universally variable within species. In topi, both lekking and resource-defending individuals can be found within the same population (Gosling and Petrie 1990) and populations vary in the extent of lekking (Gosling 1986). The same is true for fallow deer, where the same populations have been reported to change their most common mating system (including leks) between different years (Langbein and Thirgood 1989). Uganda kob seems to be the most obligate lekking mammal. However, another subspecies, the white-eared kob, is more plastic, while yet another subspecies, the Senegal kob, does not lek at all (Balmford 1990).

Apart from the marked intraspecific variation within ungulates, lek breeding seems very similar in most aspects when compared to classical bird leks. Territories of individuals are rather constant over time, leks occur on traditional sites, and the skew in male mating success is pronounced.

In some of the listed mammals it is still uncertain if what has been described as lek mating really conforms to a strict definition of a lek. In the case of reindeer and walrus, the information is still scanty and uncertain. Walruses probably do not lek (P. K. Anderson, pers. comm.). In the case of the wildebeest, lek mating may be inferred from the study of Estes (1968). However, Estes contrasted territoriality in wildebeest with lekking in Uganda kob (p. 296), and the spacing of territorial males seemed to be larger than needed to qualify as lekking males.

Studies of bats suggest that female home-range size becomes large when the diet of a given species consists of food that is either hard to find or distributed so that females forage in trap lines (Bradbury 1977, Bradbury and Vehrencamp 1977). In antelopes, like in the grouse, lekking occurs in populations where females have large overlapping home ranges (fig. 2.4, bottom). The general pattern is that lekking is associated with high population densities such as in fallow deer (Langbein and Thirgood 1989), Uganda kob (Balmford 1990), and in sika deer (Balmford et al. 1993). It seems that high-density situations decrease the costs for females to visit clumped male arenas.

Male emancipation is much more common in mammals than in birds, and yet lekking is relatively rare in mammals. A possible explanation for the rarity of lekking is the limited mobility of mammals in comparison to birds. Clumping in traditional sites might enhance the risk of predation more often than in birds, and, in particular, in low density populations the costs for female mammals to visit the leks might be high.

2.11 Summary

Lekking has evolved independently in a wide variety of taxa. Since lekking was first described in birds and the first case studies were avian, this group is the best known. In other taxa, mating aggregations have not been called leks until recently. Thus, considerable confusion arises when inferences are drawn in other taxa, with the result that many possible cases are overlooked. Leks in mammals appear most similar to leks in birds: both show territoriality, have traditional mating sites, and exhibit pronounced variance in male mating success. Other groups commonly differ in one or more of these aspects.

We have listed some basic prerequisites for lekking to evolve. Male emancipation from any parental duties is necessary, thus paternal care limits lek evolution in many altricial birds and in many fishes. Males should not be able to control essential resources for females, a factor that is variable within all the major taxonomic groups. External fertilization constrains the evolution of lekking in most fishes and amphibia because the site of fertilization is frequently at the site of mate choice. High mobility enhances lekking in birds, fish, and some insects, while lack of mobility leading to high costs for females to visit leks and high costs of predation at lekking sites constrains lekking in other groups. Limited possibility of females to be choosy constrains lekking in insects where mating aggregations are frequently close to or at resources.

Table 2.A

A List of Lekking Taxa

Taxa		Source
ARTHROPODA		
Insecta		
Anisoptera		
Plathemis lydia		Campanella & Wolf 1973
Coleoptera		
Luciola obsoleta		Lloyd 1972
L. cruciata	Gengi firefly	Kiichiro 1961 in Lloyd 1979
Pteroptyx spp	Folded-wing fireflies	Lloyd 1979
Diptera		
Gasterophilus intestinalis		Ringo 1976
Empis borealis		Svensson & Petersson 1987
Drosophila neogrimshawi	Adiastola subgroup	Spieth 1978
D. ornata	"	Spieth 1978
D. touchardie	"	Spieth 1978
D. adiastola	"	Spieth 1978
D. cilifera	"	Spieth 1978
D. clavisetae	"	Spieth 1978
D. hamifera	"	Spieth 1978
D. paenehamifera	"	Spieth 1978
D. peniculipedis	"	Spieth 1978
D. spectabilis	"	Spieth 1978
D. truncipenna	"	Spieth 1978
D. varipennis	"	Spieth 1978
D. ochrobasis	"	Spieth 1978
D. setosimentum	"	Spieth 1978
D. cnecopleura	Planitibia subgroup	Shelly 1988
D. conformis	"	Shelly 1987
D. heteroneura	"	Spieth 1981
D. silvestris	"	Spieth 1981
D. setosifrons	"	Spieth 1978
D. planitibia	"	Spieth 1978
D. obscuripes	"	Spieth 1978
D. neoperkinsi	"	Spieth 1978
D. melanocephala	"	Spieth 1978
D. ingens	"	Spieth 1978
D. hanaulae	"	Spieth 1978
D. differens	"	Spieth 1978
D. cyrtoloma	"	Spieth 1978
D. oahuensis	"	Spieth 1978
D. nigribasis	"	Spieth 1978
D. hemipeza	"	Spieth 1978

Table 2.A *(cont.)*

A List of Lekking Taxa

Taxa		Source
ARTHROPODA, Diptera *(cont.)*		
D. pictiornis	"	Spieth 1978
D. affinidisjuncta	Grimshawi subgroup	Spieth 1978
D. balioptera	"	Spieth 1978
D. bostrycha	"	Spieth 1978
D. crucigera	"	Spieth 1978
D. disjuncta	"	Spieth 1978
D. grimshawi	"	Hodosh et al. 1979
D. pullipes	"	Spieth 1978
D. mycetophaga[a]	Hirtodrosophila	Parsons 1978
D. polypori	"	Parsons 1977a,b
D. mixtura	"	Parsons 1977a,b
Anastrepha ludens		Aluja et al. 1983
A. oblicua		Aluja et al. 1983
A. suspensa		Dodson 1982
A. fraterculus		Morgante et al. 1983
Zygothrica dispar		Burla 1990
Procecidochares sp.		Dodson 1986
Dacus cucurbitae	Melon fly	Iwahashi & Majima 1986
D. longistylus		Hendrichs & Reyes 1987
D. dorsalis		Shelly & Kaneshiro 1991
Ceratitis capitata		Arita & Kaneshiro 1985
Euaresta bella		Batra 1979
E. festiva		Batra 1979
Physiphora demandata		Alcock & Pyle 1979
Lepidoptera		
Estigmene acrea		Willis & Birch 1982
Papilio zelicaon		Shields 1967
P. polyxenes	Black swallowtail	Lederhouse 1982
Atlides halesus		Alcock 1983
Oenis chryxus		Knapton 1985
Perrhybris pyrrha		DeVries 1978
Coenonympha pamphilus	Small heath butterfly	Wickman et al., in press
Hymenoptera		
Hemipepsis ustulata	Tarantula hawk wasp	Alcock 1981
Xylocopa varipuncta		Alcock et al. 1987
Eulaema meriana		Kimsey 1980
Euglossa imperialis		Kimsey 1980
Centris adani		Thornhill & Alcock 1983
Polistes dominulus		Beani & Turilazzi 1990
Eucerceris flavocincta		Steiner 1978

Table 2.A *(cont.)*

A List of Lekking Taxa

Taxa		Source
CHORDATA		
Osteichthyes		
Cyprinodontiformes		
Poeciliidae		
Poeciliopsis occidentalis	Gila topminnow	Constantz 1975
Cyprinidae		
Cyprinodon sp.		Kodrick-Brown 1977
Perciformes		
Labridae		
Halichoeres melanochir		Moyer & Yogo 1982
Scarus vetula	Queen parrotfish	Clavijo 1983
Cichlidae		
Cyrtocara eucinostomus		McKaye et al. 1990
C. argyrosoma		McKaye 1983
C. nigritaeniatus		McKaye 1983
Lethrinops furcicauda		McKaye 1983
L. aurita		McKaye 1983
L. variabilis		McKaye 1983
L. liturus		McKaye 1983
Tilapia grahami		Coe 1966
T. karomo		Lowe 1952
T. variabilis		Lowe-McConnell 1956
T. leucosticta		Lowe-McConnell 1957
T. nilotica		Lowe-McConnell 1958
T. machrochir		Ruwet 1962
T. melanopleura		Ruwet 1962
Haplochromis desfontainesi		Kirschofer 1953
H. heterodon		Fryer & Iles 1972
H. encinostomus		Fryer & Iles 1972
H. chrysonotus		Fryer & Iles 1972
H. burtoni		Fernald & Hirata 1977
Oreochromis mossambicus		Bowen 1984
AMPHIBIA		
Anura		
Ranidae		
Rana catesbeiana	Bullfrog	Emlen 1976
Bufonidae		
Bufo calamita	Natterjack toad	Arak 1982
B. viridis	Green toad	J. Höglund, unpubl.
B. woodhousei	Woodhouse's toad	Sullivan 1982
B. rangeri	Raucous toad	Cherry 1993
B. pardalis	Leopard toad	Cherry 1992

Table 2.A *(cont.)*
A List of Lekking Taxa

Taxa		Source
AMPHIBIA, Anura *(cont.)*		
Leptodactylidae		
Uperoleia laevigata[b]	Red-groined toadlet	Robertson 1990
Physalaemus pustulosus	Túngara frog	Ryan 1983
Hylidae		
Ololyglon rubra		Bourne 1992
Hyla versicolor	Grey treefrog	Sullivan & Hinshaw 1992
Urodela=Caudata		
Salamandridae		
Triturus cristatus	Crested newt	Hedlund & Robertson 1989
MAMMALIA		
Marsupialia		
Dasyuridae		
Antechinus stuartii	Brown antechinus	Lazenby-Cohen & Cockburn 1988
Chiroptera		
Pteropodidae		
Hypsignathus monstrosus	Hammerheaded bat	Bradbury 1977
Pinnipedia		
Odobenidae		
Odobenus rosmarus[c]	Walrus	Fay et al. 1984
Artiodactyla		
Cervidae		
Cervus dama	Fallow deer	Pemberton & Balmford 1987
C. nippon	Sika deer	Balmford et al. 1993
C. elaphus	Red deer	J. Carranza, pers. comm.
Rangifer tarandus[c]	Reindeer	Leader-Williams 1988
Bovidae		
Kobus kob thomasi	Uganda kob	Buechner & Schloeth 1965
K. kob leucotis	White-eared kob	Fryxell 1987
K. leche	Lechwe	Robinette & Child 1964
Connochaetes taurinus[c]	Blue wildebeest	Estes 1968
Damaliscus lunatum	Topi	Gosling et al. 1987
Sirenia		
Dugong dugong	Dugong	P. K. Anderson, pers. comm.
AVES		
Galliformes		
Phasianidae		
Tetraoninae		
Centrocercus urophasianus	Sage grouse	Wiley 1973
Dendragapus obscurus[c]	Blue grouse	Lewis 1985
Tetrao urogallus	Capercaillie	Hjorth 1970
T. parvirostris	Black-billed capercaillie	Hjorth 1970

Table 2.A *(cont.)*

A List of Lekking Taxa

Taxa		Source
AVES, Tetraoninae *(cont.)*		
T. tetrix	Black grouse	Hjorth 1970
T. mlokosiewiczi	Caucasian black grouse	Hjorth 1970
Tympanuchus cupido	Greater prairie chicken	Hjorth 1970
T. pallidicinctus	Lesser prairie chicken	Hjorth 1970
T. phasianellus	Sharp-tailed grouse	Hjorth 1970
Phasianinae		
Argusianus argus	Great argus pheasant	Davison 1981
Pavo cristatus	Peafowl	Petrie et al. 1991
Meleagrinae		
Meleagris gallopavo[d]	Turkey	Watts 1968
Gruiformes		
Otidae		
Otis tarda	Great bustard	Gewalt 1959
Charadriiformes		
Scolopacidae		
Philomachus pugnax	Ruff	Hogan-Warburg 1966
Tryngites subruficollis	Buff-breasted sandpiper	Myers 1979
Gallinago media	Great snipe	Lemnell 1978
Psittaciformes		
Psittacidae		
Strigops habroptilus	Kakapo	Merton et al. 1984
Apodiformes		
Trochilidae		
Glaucis hirsuta	Rufous-breasted hermit	B. Snow 1973
Threnetes ruckeri	Band-tailed barbthroat	Stiles & Wolf 1979
Phaethornis ruber	Reddish hermit	B. Snow 1973
P. guy	Green hermit	B. Snow 1974
P. superciliosus	Long-tailed hermit	Stiles & Wolf 1979
P. longuemareus	Little hermit	Wiley 1971
Eutoxeres aquila	White-tipped sicklebill	Stiles & Wolf 1979
Phaeochroa cuvierii	Scaly-breasted hummingbird	Wolf 1970
Campylopterus hemileucurus	Violet sabrewing	Stiles & Wolf 1979
Klais guimeti	Violet-headed hummingbird	Payne 1984
Amazilia amabilis	Blue-chested hummingbird	Stiles & Wolf 1979
Topaza pella	Crimson topaz	Davis 1958
Calypte anna	Anna's hummingbird	Stiles 1973
Selasphorus platycercus	Broad-tailed hummingbird	Barash 1972
Piciformes		
Indicatoridae[c]		
Indicator indicator	Black-throated honeyguide	Ranger 1955
I. minor	Lesser honeyguide	Ranger 1955
I. variegatus	Scaly-throated honeyguide	Ranger 1955

Table 2.A *(cont.)*

A List of Lekking Taxa

Taxa		Source
AVES *(cont.)*		
Passeriformes		
Tyrannidae		
Mionectes oleagineus	Ochre-bellied flycatcher	Lill 1976
M. macconnelli	McConnell's flycatcher	B. Snow & Snow 1979
M. rufiventris	Grey-hooded flycatcher	B. Snow & Snow 1979
Pipridae		
Machaeropterus pyrocephalus	Fiery-capped manakin	Robbins 1985
M. regulus	Striped manakin	de Schauensee et al. 1978
M. deliciosus	Club-winged manakin	Willis 1966
Pipra aureola	Crimson-hooded manakin	D. Snow 1963a
P. fasciicauda	Band-tailed manakin	Robbins 1983
P. erythrocephala	Golden-headed manakin	Lill 1976
P. mentalis	Red-capped manakin	Skutch 1969
P. coronata	Blue-crowned manakin	Skutch 1969
P. chloromeros	Round-tailed manakin	Robbins 1985
P. filicauda	Wire-tailed manakin	Schwartz & Snow 1978
Manacus manacus	White-bearded manakin	Lill 1974
M. manacus vitellinus	Golden-collared manakin	Willis & Eisenmann 1979
Chiroxiphia caudata	Swallow-tailed manakin	Foster 1981
C. pareola	Blue-backed manakin	D. Snow 1963b
C. linearis	Long-tailed manakin	Foster 1977
Corapipo gutturalis	White-throated manakin	Davis 1949
C. leucorrhoa	White-ruffed manakin	D. Snow 1963c
Ilicura militaris	Pin-tailed manakin	Payne 1984
Cotingidae		
Cotinga maynana	Plum-throated cotinga	D. Snow 1982
Lipaugus unirufus	Rufous piha	D. Snow 1982
L. vociferans	Screaming piha	D. Snow 1982
L. fuscocinereus	Dusky piha	D. Snow 1982
Pyroderus scutatus	Red-ruffed fruit crow	D. Snow 1982
Rupicola rupicola	Guianian cock-of-the-rock	D. Snow 1982
R. peruviana	Andean cock-of-the-rock	D. Snow 1982
Phoenicercus carnifex	Guianian red cotinga	D. Snow 1982
P. nigricollis	Black-necked red cotinga	D. Snow 1982
Procnias tricarunculata	Three-wattled bellbird	D. Snow 1982
P. alba	White bellbird	D. Snow 1982
P. averano	Bearded bellbird	D. Snow 1982
P. nudicollis	Bare-throated bellbird	D. Snow 1982
Cephalopterus glabricollis	Bare-necked umbrella bird	D. Snow 1982
C. ornatus	Amazonian umbrella bird	D. Snow 1982
C. penduliger	Long-wattled umbrella bird	D. Snow 1982
Perissocephalus tricolor	Capuchinbird	D. Snow 1982

Table 2.A *(cont.)*

A List of Lekking Taxa

Taxa		Source
AVES, Cotingidae, *(cont.)*		
Tijuca atra	Black-and-gold cotinga	D. Snow 1982
Oxyruncidae		
Oxyruncus cristatus	Sharpbill	Stiles & Whitney 1983
Pycnonotidae		
Pycnonotus latirostris	Yellow-whiskered greenbul	Brosset 1982
Ploceidae		
Euplectes jacksoni	Jackson's widowbird	van Someren 1947
Vidua macroura	Pin-tailed whydah	Shaw 1984
V. paradisaea	Paradise whydah	Payne 1984
V. orientalis	Broad-tailed paradise whydah	Payne 1984
V. orientalis obtusa	Angolan paradise whydah	Payne 1984
V. chalybeata	Village indigobird	Payne 1973
Ptilonorhynchidae[c]		
Ptilonorhyncus violaceus	Satin bowerbird	Vellenga 1970
Prionodura newtoniana	Newton's golden bowerbird	Gilliard 1969
Scenopoeetes dentirostris	Tooth-billed catbird	Cooper & Forshaw 1979
Amblyornis macgregoriae	MacGregor's bowerbird	M. Pruett-Jones & Pruett-Jones 1982
Paradisaeidae		
Semioptera wallacei	Wallace's standardwing	Beehler & Pruett-Jones 1983
Cicinnurus regius	King bird of paradise	Beehler & Pruett-Jones 1983
Pteridophora alberti	King of Saxony bird of paradise	Beehler & Pruett-Jones 1983
Paradisaea raggiana	Raggiana bird of paradise	Beehler & Pruett-Jones 1983
P. apoda	Greater bird of paradise	Beehler & Pruett-Jones 1983
P. minor	Lesser bird of paradise	Beehler & Pruett-Jones 1983
P. rubra	Red bird of paradise	Beehler & Pruett-Jones 1983
P. guilielmi	Emperor of Germany bird of par.	Beehler & Pruett-Jones 1983
P. decora	Goldie's bird of paradise	Beehler & Pruett-Jones 1983
Parotia sefilata	Arfak parotia	Beehler & Pruett-Jones 1983
P. carolae	Queen Carola's parotia	Beehler & Pruett-Jones 1983
P. lawesii	Lawes' parotia	Beehler & Pruett-Jones 1983
P. wahnesi	Wahnes' parotia	Beehler & Pruett-Jones 1983

SOURCE: The taxonomy of mammals follows Corbet and Hill (1991) and that of birds follows Howard and Moore (1991).

[a] Probably not a lekker (see Hoffman and Blows 1992).

[b] Formerly *U. rugosa.*

[c] Probably not a lekker.

[d] Lekking reported in only one population.

PART II   Sexual Selection

3

Determinants of Male Mating Success

3.1 Introduction

The study of sexual selection in lekking species has received considerable attention during the last decades (see reviews in Balmford 1990, Harvey and Bradbury 1991, Wiley 1991). Interest in this kind of research has to a large extent been motivated by the assumptions that female choice of particular kinds of males is the major cause of nonrandom mating in males and that the choice of females is based on male characters that are not related to immediate fitness gains such as access to resources and paternal care. In this chapter and the next we will challenge these premises—not so much to show that the premises are false, as the data for such a conclusion do not exist, but to draw attention to areas of research we feel have been neglected. Recently the assertions that female choice is the major determinant of male success, and that genetic benefits are the sole gain to discriminating females, have been challenged (chapter 4; Reynolds and Gross 1990, Kirkpatrick and Ryan 1991).

The assertion of strong selection on males and prevailing female preferences has led theoretical population biologists to coin the phrase "paradox of the lek" (Borgia 1979, Taylor and Williams 1982, Wiley 1991, Williams 1992). Strong selection on any male trait, and in particular traits closely related to male fitness, ought to diminish the additive genetic variance for such traits, leading to females in the long run having no variability to choose from (Fisher 1958, Charlesworth 1987, Kirkpatrick and Ryan 1991). The paradox is that in lekking species in which the only presumed benefit to females is the genetic quality of their mates (which according to theory should be the same among all males), females appear more discriminating than females of resource-defending species where females accrue direct benefits by being choosy (Reynolds and Gross 1990).

The relative roles of male-male competition and female choice, respectively, have been difficult to assess in lekking species. While the emerging picture seems to be that male success is determined by a multitude of traits, it is less clear whether these traits give males an advantage because of female preference or intramale competition (Gibson et al. 1991, Gratson et al. 1991, R. V. Alatalo, J. Höglund, A. Lundberg, and P. T. Rintamäki, unpubl.; chapter 5). In this chapter we will review studies that have addressed the question of which traits give lekking males a mating advantage and whether these are subjected to female preference and/or male-male competition. Regardless of whether a particular trait is beneficial in terms of a mating advantage or through male-male competition, we also review the potential costs that set an upper limit to further exaggeration of the trait.

On leks, female choice could operate on two levels: between leks and within them. In this chapter we focus on within-lek selection. The consequences for the spacing behavior of males and females of between-lek selection will be addressed in chapters 7 and 9. However, detailed studies of sage grouse have shown that different male traits may explain the variance in male mating success differently on different leks (Gibson and Bradbury 1986, Gibson et al. 1991). Possible explanations for this in sage grouse and other species will be discussed in this chapter.

Before summarizing the male traits associated with male mating success, we will first discuss the strength of sexual selection in terms of mating skew. We will also discuss lek organization with respect to male territoriality and male spacing.

3.2 Mating Skew

One of the special features of leks is the relatively strong sexual selection among males, with matings frequently being highly skewed within each aggregation. While high variance in male mating success is typical for leks, it is also common in many other polygynous mating systems. In particular in harem polygyny, where males directly control access to females, matings may be highly skewed within each breeding season. Examples of such species are the elephant seal (Cox and LeBouef 1977) and the red deer (Clutton-Brock et al. 1982). The mating skew in leks is not always clear; nevertheless, we will try to describe and understand that variation.

Many studies have described mating skews to be much steeper than what could be expected from any random distribution of matings in a single lek aggregation during a single mating season (for reviews, see Bradbury et al. 1985, Wiley 1991). In the black grouse, only about half of the territorial males receive any copulations, and nearly all the matings concentrate on only one-fifth of the males (fig. 3.1; Alatalo et al. 1992). Various kinds of parametric indices have been used to describe mating skew, but we prefer not to use such indices because of their obvious biases in relation to any variation in the number of males. The problem is the same as in the use of evenness indices for the comparison of the equitability of species abundances for communities of different species richness (see Alatalo 1981). There is no single way to define when any two leks with a different number of males have the same mating skew, and indeed we found major biases for the indices of skew used in some studies.

Assuming that skew is due to female mating preferences, unanimity in female choice seems to be particularly high in species where leks are small and male dominance relationships are clearly apparent. It may, therefore, be questionable whether females are free to choose males other than the dominant ones. Examples of such species are the long-tailed manakin (McDonald 1989, 1990) and the capuchinbird (Trail 1990). A comparison between leks of different size in the black grouse (fig. 3.1) reveals a clear tendency for higher unanimity in the smallest leks. This may be related to the clear dominance hierarchies in small leks, since it seems that in small leks the dominant male is generally able to control his access to females. Territory boundaries are less clear in small leks than in the larger leks, where most males in the presence of females are able to prevent intrusions by other males into their territories. In the black-and-white manakin, there was also a stronger mating skew in a small lek of ten males compared to a larger lek of twenty males (Lill 1974).

Apart from the reduced possibilities of dominant males to control access to females, the reduced skew on larger leks may simply reflect the increased difficulties of females to judge any differences among the males with respect to the traits they favored. Other factors that may even out male mating success include the relatively high female numbers in relation to males. The most clear examples of an operational sex ratio reducing skew come from frogs with explosive mating systems. Likewise, in insect swarms the mating males leave the swarm for copulation, and thus the relatively long handling time will affect the operational sex ratio. Female-female aggression is another possible factor that may reduce mating skew since dominant females will monopolize preferred males, forcing subordinates to mate with unpreferred males. This occurs, for instance, in gamebirds, and it may be particularly effective in the peahen (Petrie et al. 1992). Female copying (section 4.3), on the other hand, will increase the mating skew.

The display behavior of males is frequently energetically demanding, thus the favored males may become exhausted after a while, and therefore the success rank of males may change. For instance, in the black grouse the males most successful early in the mating season are less likely to mate later in the season (Höglund et al. 1990a). In ungulates, such as the Uganda kob (Balmford et al. 1992) and topi (Gosling and Petrie 1990), the top males are able to hold their territories only for a week or so before they have to leave the lek to restore their prime condition.

In species where males live for several mating seasons, the true opportunity for sexual selection should be based on lifetime mating success. In harem species with highly skewed mating success in any mating season, male mating success is typically age dependent (Cox and LeBouef 1977, Clutton-Brock et al. 1982). Surprisingly, only a few studies have attempted to follow the mating success of males over their lifetime. In the black

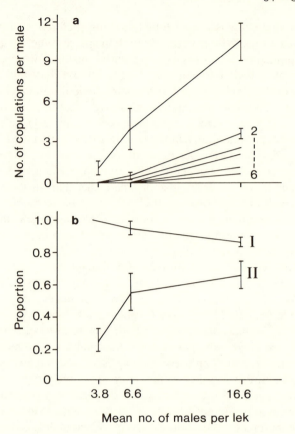

Figure 3.1 The distribution of copulations within black grouse leks of three different size categories (< 5, 5–10, > 10 males). (a) The mean number of copulations per male (± SE) for the males, grouped according to rank of mating success within each lek. In the largest lek, up to nine males received copulations, but for clarity only the six most successful males are shown. (b I) Proportion of copulations to the top-ranking one-fifth of males ($r_s = -.74$). (b II) Proportion of adult males that succeeded in copulating ($r_s = .56$). (From Alatalo et al. 1992)

grouse, the mating success increases very clearly until the age of 3–4 years, whereafter the success may even decrease (Kruijt and de Vos 1988). However, the mating success of males is not solely due to an age effect, but to other consistent differences among males (see section 3.3).

3.3 Traits Beneficial for Males

At first sight it may seem an easy task to pin down the traits that are correlated with male mating success in lekking species. Simply find a large enough lek, catch, mark and measure the morphology of all males, and quantify their behavioral differences and mating success. Then perform the appropriate multivariate statistics, and the traits that are correlated with mating success should be revealed. Although a number of studies have done this, it has turned out that one or a few single traits commonly do not explain much of the variance in male mating success. Additionally, it may be that female choices are not always so clear-cut and may be influenced by female copying. In any case, correlations alone cannot tell us which male traits are preferred by females; only experiments can reveal the traits that truly influence male mating success. So far, only very few experiments have been attempted on lekking animals.

CALLS AND VOCALIZATIONS

Spectral call properties are here defined as aspects of calls that show no or little temporal variation, for example, dominant frequency, within call durations and frequency sweeps. Such properties grade into call parameters that an individual can alter without changing the meaning of the call, such as call repetition rate. Call or display rates are here treated separately as behavioral properties. Clearly this distinction runs into problems when call parameters such as repertoire size and versatility are studied. However, we want to treat spectral call properties separately because the selection on spectral features probably acts directly on the structure of the call itself in parallel to, for example, selection on metric traits. Selection on behavioral call properties such as repetition rate does not change the spectral properties of the call and is better treated as an aspect of the behavior and not as the morphology of the animal in question.

Regardless of the mating system, spectral features of male mating calls in frogs and toads have been suggested to have evolved through sexual selection (Halliday 1983, Ryan 1991a). Two well-studied species, both with a lek-like mating system, have in particular shown that male mating success correlates with spectral call properties. In túngara frogs, males that add chucks to the whine part of their call have a mating advantage over males who do not (Ryan 1980a, 1983, 1985). This advantage is due to female preference of calls with the added chucks (Ryan 1980a). The neurophysiological basis for this preference has been investigated in some detail. It is proposed that males exploit preexisting biases in the female sensory system by adding chucks to their calls. The basilar papilla in the female ear is tuned to frequencies that match the chuck, and not only *Physalaemus pustulosus* females but also conspecific *P. coloradorum* females prefer calls with chucks. Male *P. coloradorum* do not add chucks to their calls and yet females of this species prefer them (Ryan et al. 1990, Ryan 1991b). Ryan and co-workers suggest that sexual selection by sensory exploitation can explain the evolution of male epigamic traits not only in túngara frogs

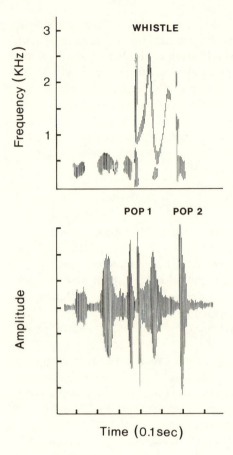

Figure 3.2 Sonogram and power spectrum of a sage grouse call. Interpop interval is the time difference between pops 1 and 2. (After Gibson et al. 1991)

but also in other species (see section 4.2 and Ryan 1991a for lists of possible examples).

The red-groined toadlet studied by Robertson (1984, 1986a,b, 1990) is another example of a species in which females prefer spectral call properties. In this case the preference seems to be based on the fundamental frequency of the male call that is correlated with male body mass. In playback experiments females preferred male calls that corresponded to a male of approximately 70% of their own mass. The suggested reason for why females prefer a certain size of males is that smaller males could not fertilize all the female's eggs and larger males drown the females (Robertson 1990). Thus there is a direct benefit for females to choose males of certain size, and in contrast to many other studies, female preference is not simply directed toward male trait exaggeration.

In sage grouse and black grouse, a probable female preference of certain spectral call aspects has been suggested. Sage grouse females seem to prefer calls that have a small inter-pop interval (Gibson and Bradbury 1985, Gibson et al. 1991) (fig. 3.2). In black grouse, successful males had shorter rookoo phrase durations than unsuccessful ones, so females seem to prefer males who complete their call in a shorter time (Kruijt and de Vos 1988). However, this rate is related to male display activity, since if males are less active they typically interrupt the initial phase of each rookoo phrase several times.

Morphological Traits

A morphological trait is here broadly defined as any metric-size trait, coloration, and the chemical composition of pheromones (table 3.1). Studies of four species of birds have produced evidence in favor of an advantage for males with larger metric traits that represent sexual ornaments. The number of eyespots in the train of peacocks (fig. 3.3, left) and the length of the head wires in Lawe's parotia were both positively correlated with male mating success (Petrie et al. 1991, Pruett-Jones and Pruett-Jones 1991). Experimental evidence shows that female Jackson's widowbirds visiting leks prefer to mate with males with longer tails (S. Andersson 1992), as in the close relative and nonlekking long-tailed widowbird (M. Andersson 1982a). Also, correlational data suggest that the probability of mating when females are already present on male territories increases with tail length (fig. 3.3, right), indicating that tail length is a short-distance cue (S. Andersson 1989). However, in an experiment elongated males also received more visits, which questions the latter interpretation. Male tail length was also shown to correlate with male condition, suggesting that females ultimately prefer such males (S. Andersson 1992).

Female great snipe seem to prefer males with larger white tail spots. Observational data have suggested that males with more white are more successful. A measure of whiteness (the number of white tail feathers with more than 50% white) showed a tendency to be correlated with male success (Höglund and Lundberg 1987). In this study the measure of success was a composite of visitation rate and copulatory success. Also, tail whiteness is the only character in which males are larger in an otherwise reversed sexually size-dimorphic species (Höglund et al. 1990b).

To experimentally investigate the effect of tail whiteness on male copulatory success, males were caught in mist nets at the onset of the breeding season and randomly assigned to one of two experimental treatments. In manipulated males the whiteness of the tail feathers was increased with Tipp-ex™, and in control birds the dark parts of the tails were filled in with dark color without altering the amount of white. During the subsequent breeding period, the visits of females to male territories and the number of copulations were recorded.

Table 3.1

Examples of Characters That Signify Successful Males in Lekking Species

Species	Trait	Correlational Evidence	Experimental Evidence	Source
MORPHOLOGY				
Jackson's widowbird	tail length	yes	yes	S. Andersson 1989, 1992
Great snipe	tail whiteness	yes	yes	Höglund & Lundberg 1987, Höglund et al. 1990a
Pea fowl	no. of eyespots	yes	yes	Petrie et al. 1991, Petrie & Halliday, in prep.
Lawe's parotia	length of head wires	yes		S. Pruett-Jones & Pruett-Jones 1991
Túngara frog	body size	yes		Ryan 1983
Fallow deer	antler size	yes		Clutton-Brock et al. 1988
"	body size	yes		Clutton-Brock et al. 1988
Uganda kob	body size and condition	yes		Balmford et al. 1992
CALLS				
Sage grouse	spectral call parameters	yes		Gibson et al. 1991, Gibson & Bradbury 1985
Black grouse	spectral call parameters	yes		Kruijt and deVos 1988
Túngara frog	spectral call parameters	yes	yes	Ryan 1980a
Red-groined toadlet	spectral call parameters	yes	yes	Robertson 1990
TERRITORY FEATURES				
Drosophila conformis	lower leaves	yes		Shelly 1987
Fallow deer	proximity to female entrance	yes		Appolonio et al. 1989a
Fallow deer	central territory	yes		Clutton-Brock et al. 1988
Topi	central territory	yes		Gosling & Petrie 1990
Uganda kob	distance to thickets	yes		Deutsch & Weeks 1992
"	popular territories	yes	yes	Deutsch & Nefdt 1992, Balmford et al. 1992
"	central territory	yes		Balmford et al. 1992
Kafue lechwe	popular territories		yes	Deutsch & Nefdt 1992
Sage grouse	dense clustered males	yes		Wiley 1973
Black grouse	central territory	yes		Kruijt & Hogan 1967, de Vos 1983
Black grouse	dense clustered males		yes	Kruijt & de Vos 1988
Sharp-tailed grouse	central territory	yes		Gratson et al. 1991
Great snipe	central territory	yes	ruled out	Höglund & Lundberg 1987, Höglund & Robertson 1990b
Ruff	central territory	yes		Shepard 1976, van Rhijn 1991

Table 3.1 *(cont.)*

Examples of Characters That Signify Successful Males in Lekking Species

Species	Trait	Correlational Evidence	Experimental Evidence	Source
BEHAVIOR				
Great snipe	display rate	yes		Höglund & Lundberg 1987, Höglund et al. 1990a
Sage grouse	display rate	yes	yes	Gibson & Bradbury 1985, 1986, Gibson 1989, Hartzler & Jenni 1988
Fallow deer	display rate	yes		Clutton-Brock et al. 1988
Woodhouse's toad	display rate	yes		Sullivan 1983
Long-tailed manakins	display rate	yes		McDonald 1989
Long-tailed manakins	quality of dual male dance	yes		McDonald 1989
Jackson's widowbird	display rate	yes		S. Andersson 1989
Ruff	display rate	yes		Shepard 1976, but see W. Hill 1991
Cichlid fish	bower size	yes		McCay et al. 1990
Several bowerbirds	bower quality	yes		Borgia et al. 1985
Lawe's parotia	display probability + rate	yes		Pruett-Jones & Pruett-Jones 1991
Fallow deer	attendance at lek	yes		Appolonio et al. 1989a,b
Ruff	attendance at lek	yes		W. Hill 1991; own obs., in prep.
Village indigobird	attendance at lek	yes		Payne & Payne 1977
Black grouse	fighting success	yes	yes	Alatalo et al. 1991
Drosophila grimsahwi	time at lek	yes		Droney 1992
Grey treefrog	time at lek	yes		Sullivan & Hinshaw 1992
"	call rate and duration		yes	"
Uganda kob	whistle bout rate	yes		Balmford et al. 1992
"	fighting rate	yes		"

The results of this experiment failed to find any evidence that tail whiteness was important in male-male competition (Höglund et al. 1990a). Instead, sixteen of seventeen matings were by males whose tails had been whitened. These sixteen matings were performed by only four whitened males, one of which obtained as many as ten of the matings. It is thus possible that these whitened males may have been attractive to females for some unknown and unmeasured reason. However, they did not differ from the rest of the males in display rate, which seemed to be the other important factor in this species.

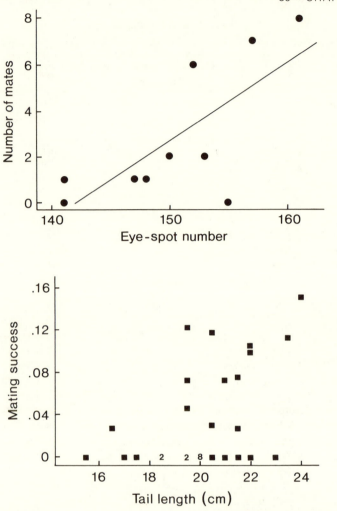

Figure 3.3 (*Top*) Relationship between peacock mating success and the number of eye-spots in the male's train (from Petrie et al. 1991). (*Bottom*) Mating success (copulations per hour) of Jackson's widowbird males plotted against tail length. Numbers indicate overlapping observations (from S. Andersson 1989).

The data can, however, be analyzed in a more powerful way by comparing the number of female visits to male territories and male mating success using a pairwise design. The territorial position of each manipulated male was determined by repeated observations and drawn onto maps with the help of a grid system (5 × 5 m) on the leks. For each manipulated male, the nearest neighboring unwhitened control male was assigned as the other part of the pair. If two or more control males were equidistant from a ma-

Table 3.2

The Mean Number of Female Visits and Copulations (±1 S.E.)
for Whitened and Control Great Snipe Males

| | Mean Values | | Paired t | P |
	Whitened	Control		
Female visits	2.0 ± 1.2	0.5 ± 1.4	1.4	.09
Copulations	0.8 ± 0.5	0.1 ± 0.1	1.9	.04

NOTES: $N=12$. Analyses were performed on $ln +1$ transformed values.

nipulated male, one of these was randomly chosen. Thus, twelve pairs of males were identified. This design should randomize any biases with respect to territorial position, differences in display rate, and other unknown factors that could also influence female choice. Test statistics are one-tailed since the hypothesis being tested is directed (more white is a priori expected to increase but not decrease mating success).

The results of the pairwise analysis are shown in table 3.2. The manipulation failed to show a statistical difference in female visitation rate. However, manipulated males showed an increased copulatory success. Thus, tail whiteness appeared to be a short-distance cue since a statistical effect was evident only for copulation success. Other factors, such as display rate, may be more important in determining female visitation rate.

In animals with large variation in body size, body mass usually is positively correlated with mating success of the males. In túngara frogs, probability of mating increases with male size (fig. 3.4; Ryan 1983). This is a common observation for amphibia, where spectral parameters are associated with body size. Also, in insects, male size is essential (table 3.1), as illustrated by the hilltopping *Hemipepsis ustulata*, where larger males occupy the highest points along the ridges (fig. 3.5; Alcock 1981). In Hawaiian *Drosophila conformis* leks, larger males tend to occupy the lowest leaves of *Pisonia umbellifera* trees, and it is these lowest parts of the lek that are favored by females (Shelly 1987).

In swarming insects, females have been found to mate with larger males that seem to occupy the sites in the swarm favored by the incoming females (Thornhill 1980, Petersson 1989). In the lovebug the size (thoracic length) of males in the bottom section of the swarm was, on average, 2.1 mm, in the middle 1.8 mm, and on top 1.5 mm (Thornhill 1980). In copulating pairs the males were, on average 2.1 mm in size while the average size of all hovering males was 1.8 mm.

In insects and frogs, male size is likely to reflect the conditions during the growth of juvenile stages. However, this does not eliminate the possibility that genetic variability also partly contributes to male size. Males that

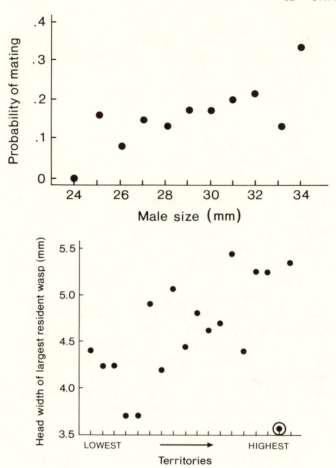

Figure 3.4 *(Top)* The frequency of mating in tùngara frog against male size. (From Ryan 1983)

Figure 3.5 *(Bottom)* Size of the "hilltopping" male tarantula hawk wasp and the territory site along the ascending ridge. Males prefer trees on higher points, and the circled outlier is for a male that defended a very small tree that was too tiny to be attractive to most males. (From Alcock 1981, and Thornhill and Alcock 1983)

are largest are the ones that have succeeded best in attaining resources during their growth in the population and season in question. If females are to assure that their offspring will inherit the genes that are locally and temporally best for offspring growth, the only thing females can do is pick the largest males. However, it may also be the case that some possible direct benefits make females choose the largest males, and at this point we can only wait for future studies to reveal the fitness consequences of females that choose larger males. The same is true for the preference of larger ornaments in birds.

In Uganda kob, body size has been shown to correlate with territorial acquisition (Balmford et al. 1992). Furthermore, in this species both body size and condition (mass corrected for size) correlated directly with male mating success when animals present on the leks were compared. Likewise, in fallow deer, mating success was correlated with large antler size and body size (Clutton-Brock et al. 1988).

Spatial and Territorial Features

In many lek species male mating success correlates with the position of the territories on the lek. In particular, the relatively higher success of central territories has been documented in many studies (table 3.1). This suggests that males compete for central positions and therefore provide females with a cue to choose the "best" males (e.g., Alexander 1975). However, fewer studies have addressed the causation behind the correlation between success and central location. The reason for the higher success rate of central territories is unlikely to be unraveled unless manipulative experiments are performed. Below we will review hypotheses and data that bear on this question, but first we will discuss studies that give possible reasons for the higher success in some territories.

In fallow deer in San Rossore, Italy, one lek was situated in a forest clearing along one of the main routes used by females (fig. 3.6; Appolonio et al. 1989a). The lek territories of the males were spaced out in this funnel-like clearing, and the territories closest to the favored female entrance were the most successful ones regardless of what male defended it. In this case, movement of females seemed restricted by the physical properties of the habitat, and males that obtained a territory through which many females were forced to pass clearly benefited.

Other studies suggest that the risk of predation is lower in some territories than others and therefore more females visit them. In topi, leks were found in areas of short grass where the risk of lion predation presumably was lower than in the surrounding savanna. On any specific lek, central territories were relatively safer, which may be why they were preferred by females (Gosling and Petrie 1990). Parallel findings were obtained in Uganda kob, although in this case territories far away from thickets were found to be the most popular (fig. 3.7; Deutsch and Weeks 1992). In both topi and kob, the direct relationship between safety from predation and success have not been established but were inferred from their consistent popularity by visiting females despite high male turnover in ownership.

That males inherit the success of previous owners has also been reported for lekking *Drosophila* (Shelly 1987). Furthermore, in white-eared kob it was found that males fought relatively more when popular territories were vacated (Fryxell 1987). Apart from the possibility that females prefer cer-

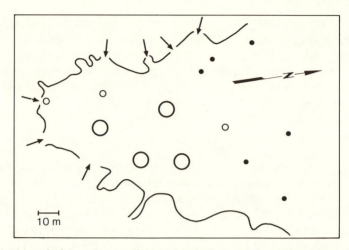

Figure 3.6 Map of a fallow deer lek. Circle size is related to male mating success and solid dots represent males without copulations. The boundary of the open area against thick scrub is indicated, and arrows refer to possible paths of entering the lek. The main path used by females was the one in the southeast corner of the opening. (From Apollonio et al. 1989)

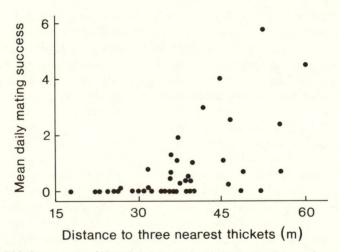

Figure 3.7 Mating success of Uganda kob males in relation to the distance from the three nearest thickets. (From Deutsch and Weeks 1992)

tain territorial features, another explanation for these findings was suggested for sage grouse. It seems that not all grouse females can mate in the territory they visit, and that not all visiting females mate on a given day but come back to mate on later mornings (Gibson 1992). It was suggested that this will create both temporal and spatial spillover. If females remember spatial features of the lek, temporal spillover would lead to females coming

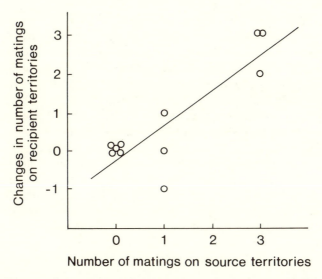

Figure 3.8 The relationship between change in mating success of lechwe territories before and after receiving earth versus the previous mating success of territories from which the earth had been taken. (From Deutsch and Nefdt 1992)

back to territories on which they saw other females on their previous visits. The timescale in which such effects can operate is not known. In black grouse, females that mate in subsequent years can in their second year mate at the same place with the same male. However, females are not faithful to specific locations, since mating success between years does not correlate when male identity has changed (Rintamäki et al. in press).

Another possibility that relates in particular to ungulates is that the odor of urine and other scent markings of preferred males or other females may still be present in the territories despite the absence of the animals who made them. By moving the topsoil from preferred to unpreferred territories in Uganda kob and Kafue lechwe, it was shown that the female visitation rates to the territories could be changed. Territories that had been unsuccessful were given soil from previously successful territories and thereby increased in popularity, and vice versa (fig. 3.8; Deutsch and Nefdt 1992). In these ungulates females seem to use the previous success of the territory as a choice criterion. This kind of temporal spillover could be seen as one way of female copying, but since successful males frequently leave their territories during the mating season and territory ownership changes, the spillover makes females mate in the same territory irrespective of who the territory owner is. Males should compete intensively for the ownership of these preferred territories, allowing females to find the most vigorous and dominant males as mating partners.

The problem with all correlational studies of traits and mating success is the intercorrelation of many factors. It is possible that males preferred by females occupy central territories for reasons unrelated to female preference for central territories. One manipulative experiment provides insight into this question. Clutton-Brock and co-workers forced the most successful male in a fallow deer lek to change position on the lek by pinning down a plastic covering over his territory. It was found that despite several removals, the male still retained his success (fig. 3.9; Clutton-Brock et al. 1989). The results of this experiment showed that an effect of male identity could be demonstrated when the effect of territory was controlled for.

Spatial spillover could be a reason for why central territories are relatively more successful than peripheral ones. If some males are more preferred than others and spatial spillover of matings occurs, it would benefit less successful males to relocate around more successful ones (Bradbury and Gibson 1983, Gibson and Bradbury 1986). Such a process would result in successful males obtaining the central territories on the lek if the number of males at the lek is large enough to completely surround those males. On very large leks multiple mating centers are predicted (Höglund and Robertson 1990b).

Removal experiments of lekking great snipe suggest that such a process is indeed operating in this species. When peripheral males were removed, their territories were instantly occupied by neighbors or nonterritorial males, whereas when central males were removed the remaining males did not occupy the vacancy that remained empty during the time of the experiment (approximately 5 hours) (Höglund and Robertson 1990b). Also, peripheral males were more likely to attack stuffed dummies accompanied by song playbacks than were central males. These results are hard to interpret if males compete for central positions as such. We suggested that this result is evidence that the position on the lek (i.e., being close to an attractive male) is more important to peripheral males than to central males. In conclusion, the relationship between central position and mating success in great snipe seems to be explained by the relocation of the less attractive males around more attractive males. Therefore in this species males are not attractive because they are central, but are central because they are attractive. Recently, Gratson and co-workers (1991) used a similar argument to explain the relatively higher success of central males in sharp-tailed grouse.

Three additional experiments suggest that males may prefer central territories. Knapton (1985) removed males in chryxus arctic butterfly leks. There was no surplus of nonterritorial males, but four out of five vacated territories in high-density areas were wholly or partially absorbed by neighboring males. All four of the territories in low-density areas remained empty, suggesting that males prefer sites with the highest density of males. In sharp-tailed grouse Rippin and Boag (1974) successively removed central males on two different leks. In both cases the remaining males readily

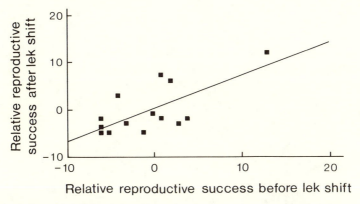

Figure 3.9 Relative reproductive success of seventeen fallow deer bucks before and after they were induced to change territories. (From Clutton-Brock et al. 1989)

took over the vacant central area, and the removal had a general centripetal effect on the male territories. A similar type of experiment on greater prairie chickens also gave indications of centripetality (Ballard and Robel 1974). However, these two experiments on grouse had no controls with the removal of peripheral males.

In insects that use landmarks, males frequently compete over certain types of habitat features (fig. 3.5), and these territorial features may be essential in female choice. Also, on the leks of the Hawaiian fruitfly *Drosophila conformis*, males fight for access to the lowest leaves in the trees they use for lekking, and females prefer such locations (Shelly 1987, 1990). Another example already mentioned in the previous section comes from insect swarms where females may prefer to enter certain sections of the swarm.

It seems that spatial and territorial features are less important in some lekking species such as the great snipe, while they might be important in other species. However, experimental evidence for the use of spatial features is scanty. In general, the use of spatial features would force the males to compete for territorial positions, allowing the females to pick the socially dominant males. Thus females would be able to choose the most vigorous males, although this could also be achieved by using morphological ornaments or behavioral activity as the criteria of choice.

BEHAVIORAL TRAITS

In contrast to morphological properties, behavioral differences between successful and unsuccessful males seem quite common in lekking species. In particular, in many studies differences in various aspects of display rate are often correlated with male mating success (table 3.1, fig. 3.10). Differ-

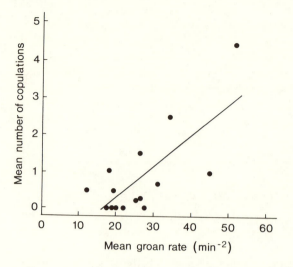

Figure 3.10 Mating success of fallow deer bucks versus their mean rate of groaning. (From Clutton-Brock et al. 1988)

ences in display rate often vary with the proximity to females on the lek; therefore the causation behind this relationship can again be questioned (Wiley 1973, 1991). However, some studies have corrected for this possible bias (e.g., Gibson et al. 1991, Hartzler and Jenni 1988) while female presence in some species may have little influence on display rate (e.g., Höglund and Lundberg 1987, Höglund et al. 1990a). Furthermore, one experiment where the display rates of males were augmented by broadcasting male song through loudspeakers showed that sage grouse females were attracted by increased display rates (Gibson 1989).

Other traits, which, like display rate, could be seen as performance cues and have been shown to be correlated with male success are the quality of the dual male dance in long-tailed manakins (McDonald 1989), the ability to build a large sand mound in a cichlid fish (McKaye et al. 1990), and bower quality in several bower birds (Borgia et al. 1985). We consider mound size and bower quality as performance cues since the selection for exaggeration of such traits must act through the abilities of the males to build them. Another question, however, is whether any of the bowerbirds can be considered to be truly lekking. At best they represent species that use landmarks without any tight clumping of males.

Another aspect of behavior that has been shown to be important is male attendance or persistence on the lek (table 3.1). While it is evident that presence on the lek must explain some of the variance in success simply because males that are present longer are exposed to more females, at least in one species, the ruff, there is some evidence that the relationship is more complicated.

The ruff is a wader that is unusual among lekking birds in that males are present on the leks almost all the time when there is enough light to see (van Rhijn 1991). This means that in Scandinavia, males arrive at the leks at approximately 3 A.M. and leave at 9 P.M. Males vary to a large extent in their presence. Some males hardly ever leave while others leave the lek for shorter or longer periods presumably to feed (Höglund et al. 1993). Male attendance rate is also highly correlated with mating success (fig. 3.11; see also Hill 1991). Female behavior is also unusual compared to other lekking birds. While in most other species both male activity and female visits are concentrated to a given time of the day, in ruffs females can, even if there is a peak of activity in the morning, arrive and mate at any time during the day. Individual females pay multiple visits to leks, which suggests that females may be assessing the presence of males. Note that we do not need to invoke consciousness or memory abilities in females to get such a process. Females could simply be selected to visit leks often and at any given time of the day. If females behave in this way it means that they force males to stay on the lek throughout the day, and males who can sustain this stay will achieve more matings.

Some studies have suggested that the rate at which individuals are involved in fights is correlated with male mating success (Floody and Arnold 1975, Trail 1985a, Appolonio et al. 1989b, Balmford 1990, Landel, in prep.). However, a problem behind such a relationship is that this could be due to correlations with other variables. For example, in Guianan cock-of-the-rock, fighting rate is highly correlated with a rank measure of fighting success (Trail 1985a). However, Trail points out that successful male cock-of-the-rock do not attack other males courting females more than the less successful males, which suggests that the higher fighting rate of successful males is the result of self-defense rather than endogenous aggressiveness.

It has long been believed that in a range of species the most dominant males are also the most successful (e.g., Ludwig 1894, Selous 1927, Brüll 1961, Scott 1942, Lumsden 1961, Koivisto 1965). However, not until recently has this belief been rigorously tested (Alatalo et al. 1991). The problem with dominance is to define the concept in an unambiguous way. Many of the studies of dominance are devoted to the establishment of stable hierarchies or peck orders that are believed to act to minimize the amount of aggression within groups (e.g., Landau 1951). In practice, dominance is often measured as the proportion of victorious fights, and then various methods are used to rank the subjects in a dominance order (e.g., Patterson 1977, Gibson and Guinness 1980, Appleby 1983, Jamieson and Craig 1987). The evidence for the existence of hierarchies on leks is somewhat ambiguous. In great snipe the rank of males appears linear (i.e., A>B, B>C, A>C instead of A>B, B>C, C>A) (Höglund and Robertson 1990b). However, the fighting success data were obtained on the lek, and males often visited by females remained in their territories while unpopular males left their territories to attack males when the successful males had female visits.

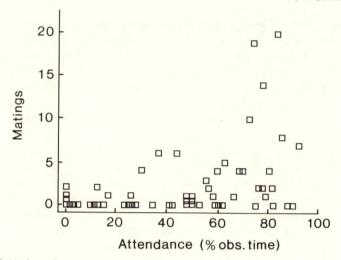

Figure 3.11 Attendance versus copulation success in male ruff (r_s = .45, N = 62, P < .002). Only males that defended a territory sometime during the breeding period are included, i.e., resident males (Hogan-Warburg 1966). Thus, marginal and satellite males are not included.

Therefore the fighting success of preferred males and the dominance hierarchy were probably inflated by ownership asymmetries.

Removal of males made the leks more unstable and increased the overall fighting rate in fallow deer (Appolonio et al. 1989b), greater prairie chickens (Robel and Ballard 1974), and golden-headed manakins (Lill 1976). One possible interpretation of these results is that the overall fighting rate increased when the hierarchy and territorial structure on the lek were destroyed. However, so far these removal experiments have been made on a very small scale without any real controls.

Because territorial males are often dominant within their territories on leks (i.e., they win almost all fights with other males), fighting success on the lek is not a useful measure of dominance (Wiley 1973, 1991). Ideally, in order to avoid biases and asymmetries due to ownership, we need to measure fighting success in nonterritorial contests away from the lek or when fights occur precisely on the border between two territories. In sage grouse, Gibson and Bradbury (1987) took advantage of an unusual situation when heavy snowfall and cold weather caused the birds to abandon the leks. In this situation males followed the females in a flock and courted them off the lek in a nonterritorial situation. Under these circumstances it was possible to obtain measurements of male fighting success, and it was shown that a measure of dominance was correlated with mating success. Our own studies of black grouse provide another opportunity to investigate the effects of dominance (see section 3.4).

Another aspect of fighting and dominance is disruption of the courtship of a pair by another male. As noted by Trail (1985a), courtship disruption has been reported in all well-studied lekking birds, and it ranges in intensity from 2% to 68% of all observed matings. Courtship disruption has been proposed as a way for males to redirect the mate choice of females in favor of males who disrupt the courtship of others (Trail 1985b). Data from the Guianan cock-of-the-rock support this hypothesis (Trail 1985b, Trail and Koutnik 1986). Trail and co-workers observed 161 instances of disrupted courtship where the identity of both the male and the female was known. They were also able to follow the subsequent female courtship and mating, the mating success of the victimized male, and the mating success of the disrupting male. Disrupted females subsequently performed more courtship visits, visited more males, and mated with more males than undisturbed females. Trail was able to observe the fate of 72 mating bouts by marked females who had been disrupted. In total these females performed 230 visits to male territories: 29 visits to males who had previously disrupted, and 201 to nondisrupters. Males that had previously disrupted were more likely to mate than nondisrupters (18 of 29 visits vs. 83 of 201 visits, $\chi^2 = 4.44$, $df = 1$, $P < .05$). This result was not due to successful males performing more disruption overall, but, as Trail argues, because some males redirected the choice of some females. In cock-of-the-rock, however, disruption is of minor importance in explaining the variance in male mating success because disruption probably entails high costs in terms of energy, risks of retaliation, and lost courtship opportunities to other females (Trail 1985b). Trail argues that overall, female choice is largely free of male control (Trail 1985a,b).

Courtship disruption is also frequent in black grouse. In Finland we have observed a mean of 17% disruption of all copulations ranging from 0% to 22% on different > sized leks (fig. 3.12a). Courtship disruption is absent from small leks and quite frequent at large leks. In contrast to the cock-of-the-rock, black grouse females do not seem to redirect their choice. We observed in only one out of sixty-three cases of observed disturbed copulations that the female continued to solicit and the disrupting male mated with the female (Alatalo et al., in prep). Most disturbed females will cease courting males and leave the lek. We had eight cases of disturbed and individually ringed females in which we could observe the mate choice in a subsequent visit, and in none of these cases did the female mate with the disrupter (Alatalo et al., in prep.). Males with intermediate success not only disrupt others' copulations but also have their own copulations disrupted (fig 3.12b,c). This is because successful males can keep other males farther away when courting females and thus prevent disruption, while unsuccessful males are rarely visited by females (Alatalo et al. 1991). Females may seek undisturbed copulations and therefore seek out dominant males. How-

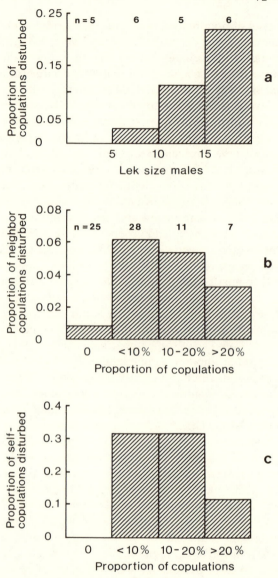

Figure 3.12 Courtship disruption in black grouse. (a) The proportion of copulations disturbed in relation to lek size. Disruption is more prevalent in larger leks (r_s = .66, N = 22 leks, 318 copulations, P < .002, overall mean disruption rate = 16.7%). (b) Proportion of neighbor copulations disrupted by self in relation to success. Males of intermediate success perform more disruptions. (c) Proportion of self-copulations disrupted by others in relation to self-success. Males of intermediate success receive more disruptions.

ever, this cannot be the entire story since black grouse females prefer to mate in larger leks (Alatalo et al. 1992) where the frequency of disturbed copulations is highest.

Furthermore, in golden-headed manakins (Lill 1976), sage grouse (Hartzler and Jenni 1988), and sharp-tailed grouse (Gratson et. al 1991), females were not more likely to remate with the disrupting male than with any other male. These results suggest that females do not redirect their mate choice in favor of disrupting males in most species studied.

AGE AND BODY SIZE

From the discussion so far of which traits are correlated with male mating success on leks, it is evident that any particular attribute of successful males is usually correlated with other characteristics. Thus the causation of events will always remain obscure in observational field studies. Even in detailed studies where multivariate statistics can be performed, it is possible that some important factor may have been overlooked. The challenge of the future is to design manipulative field experiments that control for such additional factors. However, any study has to use both approaches. Observational studies indicate which factors are correlated to male mating success, and manipulative experiments can reveal the causation behind such correlations.

Age-related mating success makes the problem of intercorrelations clear. Age could be directly correlated with mating success, while morphological, territorial, and behavioral characters that are correlated with success could also be related to age. In black grouse, for example, age is correlated with the number of copulations in a given year (de Vos 1983, Kruijt and de Vos 1988, Alatalo et al. 1991). Male success increased until the fourth year of life and thereafter showed a decline in a population in the Netherlands (fig. 3.13; Kruijt and de Vos 1988). In a subsample of males with central territories, the youngest and the oldest males had lower success than middle-aged males (Kruijt and de Vos 1988). Male black grouse who defend a central territory when they are middle-aged have higher lifetime reproductive success than males that fail to do so (fig. 3.13; Kruijt and de Vos 1988). The lower success of younger birds is probably best explained by age-dependent performance and the importance of access to a territory to attract females. The senescence of older males is harder to explain and is a phenomenon of wider evolutionary significance (Medawar 1952, Williams 1957). Among possible explanations is a pleiotropic genetic effect in which a high performance early in life leads to reduced performance later in life (Williams 1957, Gustafsson and Pärt 1990).

Age is also related to the size of the epigamic characters in black grouse. Our own studies reveal that the length of the ornamental lyre feathers in the

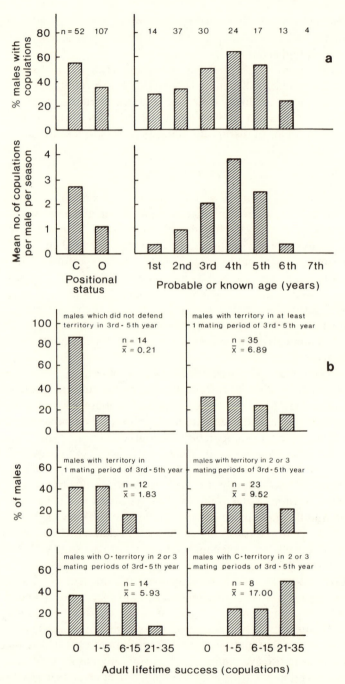

Figure 3.13 Male success in relation to age in black grouse. (a) Copulation success in relation to positional status and age. C are central territories and O other territories. (b) Lifetime copulation success of adults according to the age period in which males defended a territory and the location of the territory. (After Kruijt and de Vos 1988)

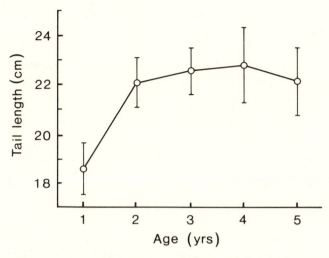

Figure 3.14 Tail length in relation to age in male black grouse. Vertical bars indicate ± 1 S.D. (ANOVA, $F_{4,160}$ = 124.3, $P < .0001$).

tail is on average 37 mm shorter in first-year males compared to older ones (first year: tail length = 18.6 ± 1.1 (S.D), $N = 94$; 2 years or older: tail length = 22.3 ± 1.1 (S.D), $N = 111$; $t = 23.4$, $P < 001$). We have no evidence of any effects of senescence in this character, although such an effect cannot be excluded (fig. 3.14). Similarly, there is a strong age effect on the size of the peacock's train (Manning 1989). Another example is the size of the white tail spots in great snipe. In a sample of recaptured males, the extent of white increased, at least up to their third year (Höglund et al. 1990a). However, this trait also shows year effects indicating that the character is conditional on environmental factors (Höglund et al. 1992a).

That characters covary with age has suggested that females may use such characters to select males that have proven their survival abilities (Trivers 1972). However, the large within-age class variation in characters shown in most examples suggests that such a process is unlikely to be particularly important. Even if characters are correlated with age, the confidence intervals commonly reported make it almost impossible to predict age from the size of the character (e.g., Alatalo et al. 1986a).

The size of epigamic characters is also commonly correlated with individual body size. The size of the antlers of fallow deer and horns of Uganda kob show such a relationship (Clutton-Brock et al. 1988, Balmford 1990), but this could simply reflect allometry. However, in a sample of both lekking and non-lekking passerines with elongated tail ornaments, all species showed positive allometry (heterogony); larger individuals had proportionally longer tails than expected from a simple scaling effect (Alatalo et al 1988). An implicit assumption behind many studies that address the question of age effects and allometric relationships is that the size of ornaments may reveal individual quality. In other words, the size of ornaments signals a good breeding situation, either for environmental or genetic reasons (Höglund, in press). The difficulty is to obtain simple measures of quality. One aspect of quality that has been suggested is the parasite load of the males.

Low Parasite Loads

Not surprisingly, recent studies have shown that parasitic infection has an adverse effect on the size and brilliance of sexually selected characters such as elongated feathers and plumage (e.g., Zuk et al. 1990a, and references in *Am. Zool. 30*(2) 1990; Loye and Zuk 1991). In lekking species, the adverse effect of parasites on tail length has long been known. Lund (1946, 1954) examined a number of phenotypic characters and gut parasite loads in capercaillie and black grouse but never analyzed the data statistically. We used Lund's data for black grouse and regressed an index of parasite burden with the length of the ornamental tail feathers and found a negative relationship (fig. 3.15; Höglund et al. 1992b). We also found that the length of the tail feathers was negatively correlated with the intensity of microfilaria found in the blood. These microfilaria were probably produced by the nematode worm *Splendidofilaria tuvensis*.

While adverse effects on ornaments seem to be commonplace, it is harder to find evidence of parasitic infection being linked to reduced male mating success. Some studies in ornamented but non-lekking animals have shown such a relationship (fish: Kennedy et al. 1987, Milinski and Bakker 1990; birds: Møller 1990, Zuk et al. 1990b,c). In all these examples, male ornaments revealed parasitic infection and females preferred males with larger ornaments. In lekking species such data are still scarce. Our studies of black grouse failed to show any effect of microfilaria load on male mating success even though male ornaments were affected (Höglund et al. 1992b). In Lawe's parotia, heavily infected males did not mate, although the relationship of parasite intensity and male mating success was still not significant (Pruett-Jones and Pruett-Jones 1990). Sage grouse in eastern California showed low levels of infection by a single hematozoan genus, *Haemoproteus*, but no measure of display performance or mating success was significantly correlated with parasite loads in adult males (Gibson 1990).

Studies of sage grouse in Wyoming showed that males infected with avian malaria (*Plasmodium pediocetii*) had lower copulation success than uninfected males, and that fewer copulations were performed by males with lice (*Lagopoecus gibsoni* and *Plasmodium pediocetii*) (Johnson and Boyce 1991). However, parasitism had no effects on strut rate, a trait previously shown to be correlated with male mating success (table 3.1), and there are no reports of effects on other possible ornaments apart from lice-

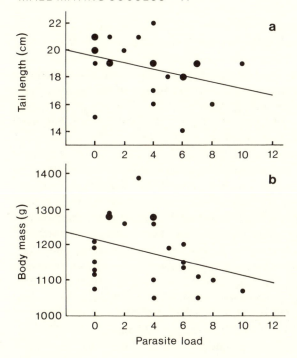

Figure 3.15 (a) Tail length in relation to gut parasite load in male black grouse ($r_s = -.58$, $N = 22$, $P < .004$, $Y = 19.6 - 0.24X$, $r^2 = .27$). (b) Weight in relation to gut parasite load in male black grouse ($r_s = -.41$, $N = 23$, $P = .05$, $Y = 1216 - 10.3X$, $r^2 = .19$). (From Höglund et al. 1992b; data from Lund 1954)

producing hematomas on the enlarged air sacs of the males that may provide females with a cue to avoid infected males.

To test if the lower success rate of parasitized males was a consequence of unattractive males being more susceptible to parasites, three experiments were performed in outdoor aviaries (Spurrier et al. 1991). In the first experiment males were treated with the antibiotic oxytetracycline to reduce overall parasite loads. Treated males did not have lower detectable loads but showed higher strut rates and were chosen more often by females than control males. In the second experiment, experimental males were dusted with carbaryl to remove lice, and in the third experiment artificial hematomas were applied using a red felt-tipped marker pen. These latter experiments showed that lice loads were reduced by the carbaryl treatment but neither strut rate nor female choice was affected. Application of artificial hematomas to males significantly reduced the number of females choosing such males, but control males also strutted more often than experimental males following the treatment.

The data on parasite loads and sexual ornaments and their possible effects on subsequent mating success are thus hard to interpret. The evidence produced so far seems to indicate that parasites do have adverse effects, but manipulative experiments, such as those by Spurrier et al. (1991), are needed to reveal the causes and discriminate between alternative hypotheses.

3.4 Male-Male Competition or Female Choice?

In the previous sections we looked at characters that give males a reproductive advantage. However, this advantage could be generated by two alternative processes. Ever since the publication of *The Descent of Man and Selection in Relation to Sex* it has been recognized that characters that give males a mating advantage could either be selected through male-male competition or by female preferences (Darwin 1871). Female choice has also been dichotomized into passive and active choice (Parker 1983, Partridge and Halliday 1984). In passive choice females simply mate more with ornamented males for statistical reasons. For example, imagine a frog species where males call in choruses. Receptive females arrive at the chorus and move toward the calling males. The louder a male calls, the further his call will be carried, allowing him a wider range of attraction. This also means that a female equidistant between two males calling at different sound-pressure levels will move toward the loudest call. Such a process does not invoke a decision process in the females. In active choice, on the other hand, the sampling and rejection of several males in favor of others is necessarily present (Parker 1983, Partridge and Halliday 1984, Gibson and Bradbury 1986). It should be noted, however, that both active and passive choice encompass selection on male attributes and thus, when considering males, constitute mechanisms that generate sexual selection (Wiley 1991).

Most authors attribute the differential mating success of males on leks to female preference, be it a passive or an active process. However, some authors have stressed the importance of male-male competition (e.g., Beehler and Foster 1988, Trail 1990). It is unfortunate that researchers tend to use the occurrence of either male-male competition or female preference as an argument that either one of the processes excludes the existence and importance of the other. While it may be true that female choice operates in many lek systems, this does not necessarily mean that male-male competition may not be operating, and vice versa.

This argument could be illustrated by the black grouse. Here, female choice undoubtedly occurs among the males present on any given lek. A female visits the lek several times before mating, moves around the lek relatively unharassed by the males, visits several males, and finally mates with one (Höglund et al. 1990b, Alatalo et al. 1991, Rintamäki et al., in prep). It should be noted that female visits to leks cannot be explained by females seeking undisturbed places, as in some ungulates (Clutton-Brock et al. 1992). Data from two leks in Sweden show that lek visits are typically short (mean = 6.7 [range 0–36] mins., $N = 138$), and during lek visits females move about (mean distance traveled during visits = 26 [range 0–112] m, $N = 138$) and visit several males (mean number of males visited = 2 (range 0–6), $N = 138$). If male coercion occurs, females typically leave the lek and forced copulations do not occur.

However, evidence of male-male competition also exists in this species. Even if no male can monopolize the entire lek and exclude other territorial males from mating within their territories, it is possible that some males occur in less favorable positions within the lek. Since the position on the lek is a factor determined by male-male competition, this process can thus generate biases among the males even before the first females have visited the lek (Rintamäki et al., in press). Furthermore, the intense competition for a position on any given lek may have excluded some males from obtaining a territory altogether and therefore minimized their reproductive possibilities (see also Gratson et al. 1991). On small leks and leks on ice-covered lakes, male-male competition is relatively more important than on large leks and on leks on bogs (M. Hovi, R. V. Alatalo, P. T. Rintamäki, unpubl.).

Territory acquisition seems to be a particularly strong process in the ruff, where approximately a third of the independent males (i.e., territory-defending behavioral morph) fail to obtain any territories at all. In contrast, among males belonging to the parasitic satellite morph that do not even attempt to acquire territories, there is less variance in male mating success (F. Widemo and J. Höglund, unpubl.).

In the ochre-bellied flycatcher, Westcott (1993) showed that several aspects of male song were correlated with the rate at which females visit male territories. It is thus likely that song in this species is subjected to intersexual selection. Westcott then examined whether song also had an intrasexual function. He therefore muted a number of males by puncturing their air sacs, making them temporarily unable to sing. He then compared male intrusion rates on muted males' territories with two groups of controls (unmanipulated and sham controls). The results showed that muted males suffered from increased levels of intrusions from other males, and five out of six muted males lost their territories while being forcibly silent. Thus, in this species, song probably has a dual function. Females are attracted to male song but male song also keeps male competitors away.

In studies of sexual selection on leks it will be necessary to tease apart the variance in male mating success and partition the observed variance to male-male competition and female choice, respectively. In this context it is important to identify which characters give males an advantage in male combat and which characters are subjected to female preference. However, if females prefer dominant males the latter would be trivial.

Female choice and male-male competition may or may not act in the same direction on any specific character. In sharp-tailed grouse it has been suggested by two independent studies that contrary to common belief, smaller males seem to be at a selective advantage in attracting females on the lek (Gratson et al. 1991, Landel, in prep.). This is probably not due to females actively assessing and selecting small-sized males. Instead, the relationship is likely to be explained by a preference for a third factor that is correlated with small male size. One possible factor could be male aggressiveness (Landel, in prep.). However, it is not known if aggressiveness and size are negatively correlated but it is possible that smaller males may be able to be more aggressive for energetic reasons. In any case, once established on the lek, smaller males seem to gain higher success than larger ones. Conversely, large size seems to give an advantage in gaining access to a territory on a lek (Gratson et al. 1991). Thus, in sharp-tailed grouse, the net selection on male size is a result of two opposing selection pressures: selection for large size through male-male competition in territory establishment, and selection for small size through female choice on the lek.

The importance of male-male competition has been especially advocated in studies of some manakins. In the genus *Chiroxiphia* males display in pairs (Foster 1977, 1981, McDonald 1989, 1990). In each pair it is predominantly the alpha male who mates while the beta male rarely has access to visiting females (McDonald 1990). Furthermore, behind each pair there may be a queue of up to eight males that have very limited access to the display perch. Copulations by such males are very rare indeed. The dominance order within these groups of males seems to be maintained by male-male competition, and since only the most dominant males mate the variance in male success within these groups is produced by competition between males. The variance generated by female preference is given by female preference for particular pairs of males, that is, between group choice.

Another example of a species where male-male competition seems to be the major factor explaining differences in male success is the capuchinbird studied by Trail (1990). In this species all copulations on a lek of eight males were performed by a single dominant male. Other examples where there is circumstantial evidence of male-male competition being more important than female choice includes the buff-breasted sandpipers, long-tailed hermits, and possibly some birds of paradise (Pruett-Jones 1988, Stiles and Wolf 1979, Le Croy 1981).

Even though the importance of male-male competition should not be dismissed in many lekking species, female preference does seem to be important. Thus many of the characters listed in table 3.1 probably give the males an advantage because females prefer to mate with males with well-developed traits. The reason for this could to a large extent be attributed to the fact that most field studies are concerned with selection within leks once leks have been established, that is, most studies concentrate on the time in the breeding cycle when females attend the lek. Also, for logistic reasons, most studies concentrate on one or a few adjacent leks during only one or a few breeding episodes. In only a few studies has more than one lek been studied and in even fewer cases have experiments been performed that exclude the effect of male-male competition for any given trait. In order to detect the importance of male-male competition, it seems important also to study the territory-establishment phase of the breeding cycle of the species in question.

In the great snipe it was possible to study the effect of experimentally enlarged white tail spots both in a male-male competition and in a female-choice context. Here it was shown that enlarged white tail spots did not give the males an advantage in fights on the lek, whereas larger white tail spots seemed to increase a male's attractiveness to females (Höglund et al. 1990a).

By placing loudspeakers broadcasting male song in sage grouse territories, Gibson (1989) was able to increase female visitation rates. This result is also better explained by female preference than male-male competition. Similarly, the tail-cutting experiments on Jackson's widowbird revealed that elongated males gained a mating advantage that is hard to attribute to differences in male dominance. Instead a female preference for elongated tails seemed to be the best explanation for this result (Andersson 1992).

A final example of a male attribute that is hard to explain as a consequence of male-male competition is the famous train of the peacock. In several recent studies it has been shown that males that have a large number of eyespots (ocelli) in the train enjoy higher mating success than males with fewer (Petrie et al. 1991). Eyespot number is correlated with train size (i.e., overall length); thus males with a large train have more eyespots. In an experiment, train length was held constant while eyespot number was manipulated. Males with reduced number of eyespots had lower mating success as compared to males with an intact number (M. Petrie and T. R. Halliday, unpubl. data). This result is not likely to be explained by male-male competition, since neither eyespot number nor train size is likely to give males an advantage in intrasexual competition. It is more plausible that a large train such as the peacock's would hamper male fighting ability.

In this context it is important to point out that we could probably search in vain for a single trait that would explain all the variance in male mating success. It could be that male-male competition and female choice favor different characters, and that selection on a trait favored by male combat may have indirect detrimental effects on the traits favored by female preference, and vice versa. However, if females are choosing the most vigorous and viable males, it is likely that females should prefer traits that are favored in male combat. In fact, females could be expected to incite male-male competition in the way suggested for elephant seals (Cox and Le-Boeuf 1977) to assure that they will mate with dominant males.

Female choice seems in most species to be based on an array of criteria. For example, in great snipe, display rate, the size of the white tail spots, and possibly other criteria seem to be important in female choice (Höglund et al. 1990a, Höglund and Robertson 1990b). In Jackson's widowbird, males with high jump display rates and with longer tails are favored (Andersson 1989, 1992). In sage grouse, the interpop interval of the call, strut rate, and characters revealing low parasite loads all seem to be of importance (Gibson et al. 1991, Spurrier et al. 1991). In short, sexual selection on leks is a complex interplay between male-male competition, female choice, and the characters favored by both processes.

3.5 Costs of Sexually Selected Traits

So far in this chapter, we have reviewed traits that may give lekking males a benefit. While the number of studies that have addressed this question are quite extensive, the number of studies that have addressed the costs of such traits are few. The costs of traits favored by sexual selection can potentially be of two forms. First, there could be energetic costs due to, for example, the growth, maintenance, or performance of a display, which would offset any further elaboration of the character. Second, any given trait that is used to attract females or signal resource holding potential to other males could also attract predators or parasites.

Studies of sage grouse have revealed that the energetic costs of male display behavior are substantial. In this species it was possible to measure the energetic expenditure of male display with the doubly labeled water technique (Vehrencamp et al. 1989). This method measures the fractional turnover of injected water labeled with low-activity isotopes (e.g., Williams and Prints 1986). The daily energy expenditure (kJ) for males with high display rates was about two times the expenditure of males that did not display at all. Furthermore, vigorously displaying males used about four times the energy required by the basal metabolic rate. Male display rate and daily energy expenditure were positively correlated (fig. 3.16).

Similar results were found in studies of great snipe. In this species the energetic costs of male display behavior were also estimated using doubly labelled water. As in sage grouse, male display rate was correlated with the estimated daily energy expenditure (Höglund et al. 1992a). Furthermore, by digging down and hiding electric balances at places where males displayed often, it was possible to measure how male body mass changed on a short-term basis. Males were found to lose on average 1.77 g/h (± 0.03 S.D., $N = 5$) of their body mass when they displayed. This corresponds to a daily mass loss of about 6.8%.

The results from both sage grouse and great snipe thus suggest that male

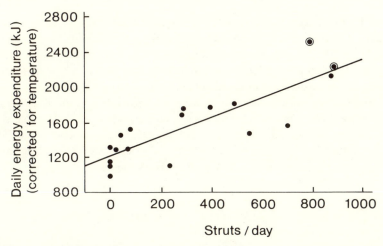

Figure 3.16 Daily energy expenditure (kJ/day) in relation to strut rate (values corrected for the effect of ambient temperature). The circled points indicate two males that were known to be successful in mating. The regression equation is $Y = 1206.3 + 1.11\ X$, adjusted $r^2 = .695$, $P = .0001$. (After Vehrencamp et al. 1989)

display rates in these species are constrained by energetic limitations. In both species the estimated daily energy expenditures were found to be about four times the basal metabolic rate. This is a value that has been suggested to be the maximum sustainable work rate for birds (Drent and Daan 1980).

Two studies of frogs have shown that calling is energetically costly (Bucher et al. 1982, Wells and Taigen 1986). In grey tree frogs, call duration and call rate were positively correlated, the call parameters explaining 84% of the variance in aerobic metabolism. Females of this species prefer metabolically costly signals and could thus potentially gain by selecting males in good physical condition (Klump and Gerhardt 1987). It may be that in this species the female preference for longer calls is open ended (i.e., that females, if they were given a choice of calls even longer than the maximum natural variation, would prefer such calls). If this is the case, it is conceivable that the energetic costs of producing long calls balance any further mating benefit. In túngara frogs, calling is also energetically expensive (Bucher et al. 1982). However, in this species it is argued that the main cost of reproduction to the males is in terms of the second type of costs—predation (Ryan 1985).

The evidence in any lekking species that predation is costly is scanty (see also chapter 6). In túngara frogs, it has been shown that predators such as bats, opossums, and other frogs use male calls to locate their prey. Calling males thus reduce their chances of survival (Ryan et. al. 1981, 1982, Ryan 1983, 1985). Thus in this species males face a trade-off where they benefit by attracting females but lose by attracting predators. Among birds, predation has been considered to be important only in cock-of-the-rock leks (Trail 1987). In African ungulates, kills by predation may play a role on leks as exemplified by the Uganda kob, where mating rates were reduced close to sites of a lion kill (Deutsch and Nefdt 1992).

Similar trade-offs have been suggested in a number of other species (Höglund, 1993). Advertising males not only attract females but also predators and/or parasites in crickets (Cade 1975, 1979, Cade and Wyatt 1984, Sakaluk and Belwood 1984), and a cicada (Soper et al. 1976). Other tentative examples of insects are reviewed by Burk (1982).

3.6 Lek Organization

The two main questions in lekking nowadays concentrate on sexual selection and the evolution of lekking itself. However, it is also important to understand how leks are organized spatially, an aspect that was frequently emphasized in the early descriptive lek studies. Here we look at the function of territoriality and the reasons for the differences in spacing of males in different species. This variation ranges from tightly clumped leks to the exploded lek.

TERRITORIALITY

Territoriality is typical for lekking systems, it is only in the aerial displays of insect swarms that territoriality is totally absent. However, even in insect swarms, males may prefer certain parts of the swarms, creating a form of relative site dependence. For instance in lovebugs, large males tend to be in the lower parts of the swarm (Thornhill 1980; see section 4.3). The absence of fixed territoriality in insect swarms makes sense since it is impossible to have boundaries without topographic features. This is neatly illustrated by the black grouse, which lack fixed territories on leks on ice-covered lakes of the northern taiga. On the ice there are no landmarks that can be used as references for territory boundaries, but the addition of landmarks will induce territories (Koivisto 1965). Without landmarks males have fixed spacing only relative to each other.

What is the function of territoriality? Do females prefer males in absolute sites or do they prefer relatively central males? Earlier, in section 3.3, we concluded that in some cases, such as insects using topographic features in mate choice, absolute sites may be important. In some other species, such as the black grouse, it is possible that relative position may have an impact, even if this is difficult to verify experimentally. In yet other species, such as the great snipe, territory position per se may not be important

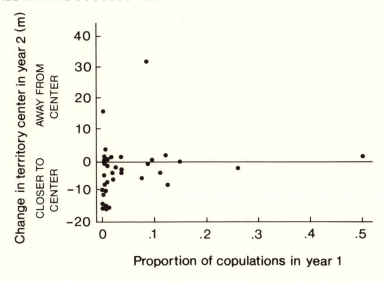

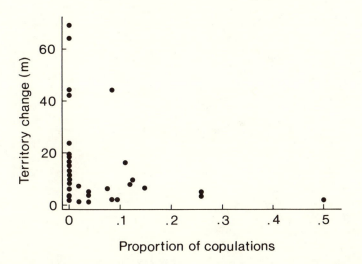

Figure 3.17 (*Top*) Change in the location of the territory center in relation to the distance to the lek center in two consecutive years for individually marked black grouse males. (*Bottom*) Change in the territory center in relation to the mating success in the previous season. (From Rintamäki et al., in press)

at all. If territory location, absolute or relative, is important for female choice, then it is obvious that males attempt to take over and defend the most popular sites. This may be so even without female preference of males in relation to the territory site, since spatial spillover will make it beneficial for males to establish territories around the male preferred by

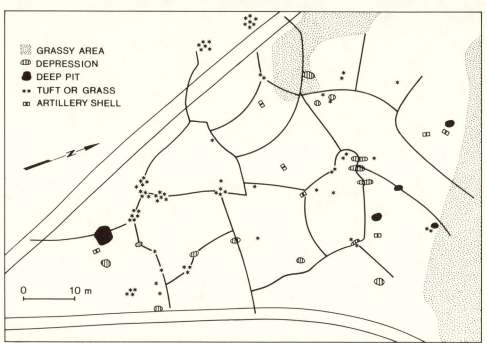

(a)

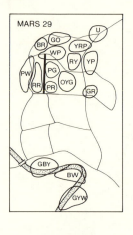

MARS 29

U
BR GO
WP   YRP
PW   PG   RY   YP
RR PR OYG
GR

GBY
BW
GYW

MAY 6

GO
BR   U   YRP
PW   PG   RY   YP
RR PR YOG
GR

GBY
BW
GYW

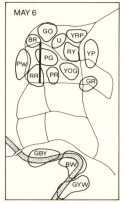

JUN 21

(BP)
BR   GO   YRP   (YRO)
(BGP)   PR   RY   YP
PW RR (U) OYG
(U)   GR (OY)

GBY
BW
(GYW)

JUL 17

YBP   RP
BGP PR   RY
(WW)   YP (U)
RR   YB   (YOG) (PP)   GR
U

GBY
(BW)

10 m

(b)

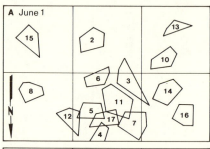

A June 1

15
2
13
10
8
6   3
11
14
12   5
17   7
4
16
N

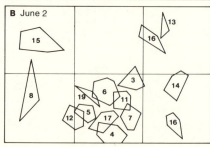

B June 2

15
13
16
8
6   3
19   11
12   5   17   7
4
14
16

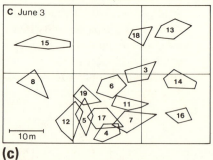

C June 3

15
18   13
8
3
6
19   14
11
12   5   17   7
4
16

10m

(c)

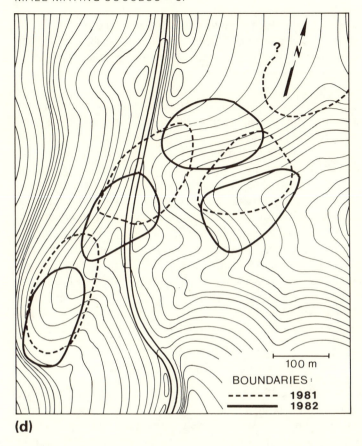

(d)

100 m

BOUNDARIES:
- - - - - - **1981**
————— **1982**

Figure 3.18 Territory maps from leks of (a) black grouse (from Hjorth 1966), (b) long-tailed hermit (Stiles and Wolf 1979), (c) chryxus arctic butterfly (Knapton 1985), (d) Costa Rican sharpbill (Stiles and Whitney 1983).

females. Another explanation for territoriality is that it is a way to minimize disturbance from other males during copulations. It is conceivable that both of these explanations are valid in many of the lekking systems.

Movement of males toward the central parts of the lek in many lekking species (section 3.3) indicates that territory location is frequently important. As an example, in black grouse, vacancies in the central parts of the lek are filled by males moving from the edges toward the center over the years (fig. 3.17, top; Rintamäki et al., in prep.), suggesting that a central position is beneficial for males even if absolute site has no effect. However, males that have once attained relatively high mating success will always stay where they are (fig. 3.17, bottom). Furthermore, males establishing new territories during the spring tend to choose sites as close to the favored males as possible, not as close to the center of the lek as possible (Rintamäki et al., in prep.). Thus spatial spillover is also likely to influence male territoriality.

The importance of territoriality in many lekking bird species is emphasized by the pattern that among resident species, males defend their territories throughout the year even if female visits and matings are limited to much shorter periods (see, e.g., Wiley 1991). In gamebirds, males typically defend their territories in the autumn directly after the moult period. In birds of paradise, territory defense extends over much of the year except for the moulting period.

In Figure 3.18 we give some examples of the spacing of male territories on leks. Black grouse (fig. 3.18a) represents classical leks with relatively small territories in relation to the size of the bird. Much of the male activity takes place at the territorial boundaries when defending the territories. In the black grouse, the distances are longer for territories at the edges of the lek and on the smallest leks. This is typical for all classical leks and reflects the impact of the increasing pressure by neighbors on the central parts of large leks. In many others, in particular in species with exploded leks, males typically concentrate their display to a fixed site(s). In fact, it may be that only such sites are defended against intruders. The long-tailed hermit hummingbird is such a species (fig 3.18b). The chryxus arctic butterfly has territories of about the same size as the black grouse and long-tailed hermit, but the size of males is much smaller (fig 3.18c). A truly exploded lek is represented by the Costa Rican sharpbill (fig 3.18d), which is a passerine of small thrush size. Territories of males are more than a hectare in size.

Spacing of Males

Why do some species have highly clumped male display territories while others have, more or less, exploded leks? Foster (1983) presented the idea that disturbance (i.e., any intrusion, courtship disruption, or challenge to the territory owner from other males) by other males forces males to have more dispersed display sites. Strictly comparable results of disturbance rates are hard to present because different authors have different ways of categorizing cases of disturbance. However, disturbance occurs both in tightly clumped leks (see section 3.5) and in exploded leks. This is not contrary to the idea that disturbance expands male spacing, since the net effect should be that males optimize their spacing so that disturbance risk becomes approximately the same in different cases. That is to say, if species with exploded leks would have tighter clumping, they would have intolerably high disturbance risk.

At least in the case of bowerbirds, wide spacing of males may be related to disturbance. In these species disturbance is particularly effective since other males may distort the bowers of others or steal objects used to decorate the bowers (Borgia et al. 1985). Some support for Foster's idea is also found in our studies of black grouse. The risk of copulation interruption by neighboring males is related to distances under 5 m between the males (fig. 3.19).

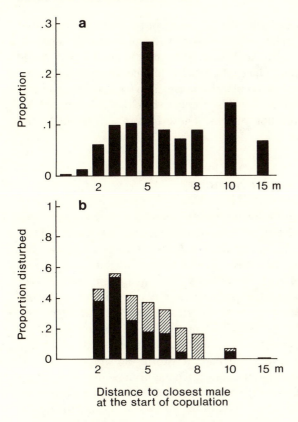

Figure 3.19 (a) Distances to closest neighboring males at the start of copulation in the black grouse. (b) The proportion of copulations disturbed (black) or under attempt of disturbance (stippled) versus the distance to the closest male. (From Alatalo et al., in prep.)

What explains the risk of disturbance for a given distance? One factor is the possibility for a disturbance to be successful. In the black grouse, 5 m is about the distance that males can run during the brief copulation. This sets a minimum for nearest-neighbor distances, given that before a copulation neighboring males typically stay at their territorial boundaries. Therefore each territory should have a radius of at least 5 m, and, indeed, this is close to the territory sizes in central parts of larger leks (see fig. 3.18a). Shorter distances between males are possible if subdominant males refrain from disturbing copulations by dominant males. This is the case in species like long-tailed manakins (McDonald 1989). On the other hand, if the effects of dominance rank are not effective, other males may disturb males courting females even before copulation. In these cases one can expect much wider spacing between the males.

Another explanation for the differences in male spacing may be related to the criteria of female choice. In many species females seem to prefer male dominance and fighting ability, and they may then refuse to mate with other than tightly clumped males. Ironically, such a female preference may involve the inability of females to mate with any other than the most dominant male of a given aggregation. However, if females were not decisive about mating in an aggregation, males would be more spaced, which would also give the subdominants a chance to mate. If females choose male ornaments or display behavior instead of dominance, then spacing may be more relaxed. Male spacing within leks is also likely to be closely linked with the factors leading to the evolution of leks. These factors will be dealt with more generally in chapter 7.

3.7 Summary

The types of traits favored by sexual selection in males vary between species, and many types of traits may be favored simultaneously. In only a few species have experiments necessary to reveal causation been performed.

Morphological attributes have been shown to correlate with male mating success in four species of birds. In addition, male mating success is positively correlated with spectral call parameters in two species of lek-breeding grouse. Body size seems to be closely associated with male mating success in insects, lower vertebrates such as frogs, and also in lekking ungulates. However, many studies fail to find any correlations with success and morphology. Spatial features of male territories are important in insects and possibly in some ungulates, but frequently not important at all in birds. The question of whether centrality of male territories has any effect is debatable, since not many experiments have been done to test the possibilities. However, experiments in great snipe indicate that the relative position, as such, need not be important at all, while manipulations in a butterfly species and some grouse suggest that centrality may play a role. Behavioral traits such as display activity and attendance at the lek are often correlated with male mating success. Experimental verification, however, is largely missing.

A central tenet of the paradox of the lek is that sexual selection in lekking species is mediated by intersexual selection. However, intrasexual selection is also common on leks and often overlooked by researchers. It is important to distinguish the mechanism whereby sexual selection is mediated. If male-male competition is the sole reason for the variance in male mating success, there is no lek paradox. However, in the great snipe, peacock and Jackson's widowbird, the importance of male morphological attributes has been verified experimentally. In all these cases male-male competition has been ruled out or is unlikely as a selective mechanism. Furthermore, experimental evidence in sage grouse suggests that display rate is favored by female preference in this species. Thus the lek paradox is still apparent, at least in these species. In the next chapter we attempt to scrutinize the second tenet of the lek paradox, that the benefit to selective lek-breeding females is indirect.

Males seem to pay relatively high costs for their displays in the few lekking species that have been examined. Assuming that such displays give males an advantage because of female preference, the preferred male traits are not arbitrary, suggesting a role for indirect female choice. Indeed, in black grouse (see chapter 5) and in the peacock, successful males are the ones that are most viable.

4

Female Mating Adaptations

4.1 Introduction

Until recently, studies of sexual selection on leks have focused on males. The question being asked has preferentially been the following: Is the variation in male mating success nonrandom, and if so are successful males in any way different from unsuccessful ones? Much attention has been directed toward identifying possible traits that identify males with high mating success. Not until recently has the interest shifted also to include variation in female behavior.

In the previous chapter we reviewed the evidence of sexual selection in lekking species. We concluded that although conclusive experimental evidence for female choice exists only in a few species, and that male-male competition is also an important mechanism in sexual selection, female preference for particular kinds of male attributes probably does exist in many lekking species. Given the lek paradox defined in the previous chapter, we may ask ourselves, why should female mate preferences evolve in lekking species?

The evolution of female mate preferences has been a contentious issue in evolutionary biology for a long time and has gained particular interest in the last decade. In this chapter we aim to review the ideas of why female mate preferences should evolve, and we will discuss the predictions and assumptions of the different models. In these models, the benefits to selective females are often assumed to be indirect, meaning that there are no immediate fitness consequences of being selective such as when the benefits are direct. Instead, females are often believed to accrue benefits through either more attractive or better-adapted offspring. Indeed, these premises of much of sexual selection theory have prompted much of the interest in leks, since on leks females have been thought only to derive indirect benefits. In this chapter we will review the alternative—that the benefits of being selective may be direct in affecting female fecundity (i.e., the number of offspring raised). Not much research has been devoted to this possibility, however, and we would like to challenge future field studies to take this possibility seriously. At the same time, we do not rule out the possibility of indirect benefits. At the present stage more field data are needed to correctly assess the possibilities.

As for any behavioral trait, female mate preferences should be amenable to a cost-benefit analysis (Reynolds and Gross 1990). If choice is costly we may expect to find adaptations in females to minimize such costs. In the same way as males trade the gains of displays and epigamic structures against the costs of expressing them, females have to trade the benefits of paying attention to such structures and comparing males against the costs of doing so. We may therefore expect to find cost-reducing strategies in selective females. One possible way to reduce costs in mate choice may simply be to copy the choice of other females. Other cost-reducing adaptations may be to use various site conventions, such as to mate on a location on the lek where the female has previously mated. Perhaps being selective among the available males does not always outweigh the costs of the choice. There are no studies that have addressed this question in lekking species, but the disappearance of female choice in the presence of predators in non-lekking sand gobies *Pomatochistus minutus* suggests that the cost-benefit analysis of choice is risk sensitive (Forsgren 1993). Such shifts between direct assessment of males and the possible prevalence of cost-reducing strategies have been suggested in sage grouse leks where the choice behavior seems to vary with the social context of the lek, in particular the number and density of visiting females (Gibson et al. 1991). Below we discuss the importance and prevalence of possible cost-reducing behaviors in mate selection and compare them with full and direct assessment of males.

Traditionally, formal models of sexual selection assume that all individuals are alike and respond equally to the selection generated by a male mating advantage. Though such models may be naive, they provide a useful baseline for investigating the dynamics of sexual selection (Kirkpatrick and Ryan 1991). Recently, several attempts have been made to incorporate differences in male quality and viability with trait expression and mating advantage (e.g., Parker 1982, Andersson 1986, Pomiankowski 1987a, 1988, Grafen 1990a,b). Alternatively, possible variation in female quality has not been considered.

Females as well as males differ in age, experience, physical condition, timing of the breeding cycle, and so on. This variation is very likely to affect their mate choice, and hence different females are likely to face different trade-offs when mating.

In most lekking bird species females seem to copulate only once (Birkhead et al. 1987, Møller and Birkhead 1991). However, in a few species some females have been observed to mate multiply both with the same and different males (e.g., Petrie in Pomiankowski 1990, Fiske and Kålås, in press). The frequency of multiple mating may also vary with female quality biases. Even though this phenomenon is poorly known, we will discuss what is known and suggest possible explanations.

Before going into the detailed discussion of the above-mentioned aspects of female mating strategies, we will outline the possible benefits that females may gain by choosing among the males on leks. These benefits must exceed the costs of visiting the leks in order for choice to evolve. As mentioned before, the benefits might be in terms of increased breeding success, and the simplest possibility for such direct benefits is that visiting the lek is the cheapest way to achieve fertilization. Alternatively or additionally, the benefits are indirect and affect the fitness of offspring.

4.2 Reasons for Female Preference

The reason females should prefer to mate with certain males and reject others has been a contentious issue for more than two hundred years (Aiken 1982). Since Darwin and Wallace, biologists have had different opinions of the underlying reasons for female preference. Darwin (1871) suggested that female preferences arose because of whims and aesthetic senses of beauty in females, whereas Wallace (1889) to a large extent disregarded female choice unless male characters revealed vigor or strength in male-male combats. These two opposing views have more or less prevailed to the present day and have in recent decades received much attention. One reason why this debate has gone on so long is probably because during the modern synthesis of evolutionary theory, sexual selection as a subject was almost completely dismissed (Huxley 1938a,b).

Since Huxley, it has taken a long time to restore the reputation and establish the importance of the field. Today most biologists agree that female preference is an important force in evolution, but, as in many other fields of evolutionary theory, the fine details of the process are still not solved. Unfortunately, empirical tests of the competing hypotheses are hard to achieve, since it is not easy to estimate the consequences of female choice on the fitness components (mating success, viability) of the offspring under natural conditions. Furthermore, it seems probable that females may be achieving many kinds of simultaneous benefits, and thus the hypotheses are complementary rather than incompatible.

This is not a book about sexual selection theory. Readers interested in this topic are advised to read recent reviews on the subject (e.g., Bradbury and Andersson 1987, Kirkpatrick 1987, Pomiankowski 1988, Harvey and Bradbury 1991, Kirkpatrick and Ryan 1991, Maynard Smith 1991, Ryan 1991b, Andersson 1994). However, since sexual selection is important not only in shaping male morphologies and behaviors on the leks but probably also is largely responsible for the evolution of the mating system, it is necessary to briefly review the thoughts on why females should show preferences for males and to state our own opinions in the matter.

FISHERIAN SEXUAL SELECTION

Fisher (1930, 1958) was the first to outline how seemingly maladaptive male traits could evolve in a population subjected to female preference. He suggested that in a population where males with certain characters gained a natural-selection advantage, females had an advantage in mating with such males since their sons would inherit the character. If the female preference and the male character later became linked so that sons would inherit the character and daughters the preference, the equilibrium expression of the trait could be moved away from the natural selection optimum. This is Fisher's famous runaway process, which is brought to a halt only when the mating advantage of the male character is balanced by increased costs due to natural selection.

Fisher never formalized this argument, but subsequent genetic models of the Fisher process have been numerous. Discrete genetic models assume two loci: a trait locus and a preference locus (e.g., O'Donald 1962, 1980, Kirkpatrick 1982). In Kirkpatrick's model, a trait locus determined the expression of the character where T males have the trait and t males lack expression; and a choice locus determined the female preference where C females mate with T males and c females mate at random. This and previous models support Fisher's general argument that female preference can spread and maintain the trait in males, given that the C trait is initially common. If, on the other hand, C females are initially rare, T males go extinct. A third outcome is possible if the initial values are about equal in number. In this case the male trait and the female preference evolve to a line of stable equilibria, and all alleles are maintained in the population (fig. 4.1). This is assuming a haploid system; diploid models are by necessity more complex and do not generate stable equilibrium lines (e.g., Heisler and Curtsinger 1990).

Quantitative genetic models reached the same general results as haploid models (Lande 1980, 1981). Depending on the genetic covariance between the male character and female preference, two general outcomes are possible. If the slope of the genetic covariance is less than the slope of the neutral line (i.e., the variance generated by sexual selection through the variance generated by natural selection), the two traits evolve toward the line of neutral equilibria. There are thus many possible combinations of male trait size and female preference. If, on the other hand, the slope of the genetic covariance exceeds the slope of the neutral line, any population will evolve away from the line. Above the line the male trait goes to fixation (Fisher's runaway process) and below the line female preference goes extinct (fig. 4.1; Arnold 1983).

There are two important points in Lande's quantitative genetic model of sexual selection. First, drift can be the initial reason for why females start

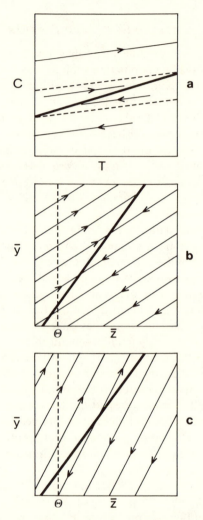

Figure 4.1 Evolutionary outcomes of Fisherian models of sexual selection. (a) Two-locus haploid model where T is frequency of the gene for male trait expression and C is gene frequency for female preference of the same trait. If initial gene frequencies lie within the dotted lines, the frequencies for male trait and female preference evolve to the equilibrium line. Above the upper dotted line, T goes to fixation, and below the line T goes extinct (after Kirkpatrick and Ryan 1991). (b) and (c) Lande's quantitative genetic model of male character $\bar{z}$ and female preference $\bar{y}$. θ denotes the natural selection optimum. When the slope of the genetic correlation of the male trait and female preference is less than the slope given by the variance in mating success divided by the variance in survival (+1), the two traits evolve toward the equilibrium line, as in (b). If the slope of the genetic correlation exceeds the equilibrium line, the preference and the trait either go extinct (below the line) or to fixation (above). The last option is the runaway case. (b and c after Arnold 1983)

to prefer males, and thus Fisher's assumption of an initial natural selection advantage is relaxed. Second, if choice is costly, the line of stable equilibria disappears. This will lead to the elimination of choosiness (Lande 1980, Pomiankowski 1988).

Quantitative genetic models of the Fisher process require a large effective population size to maintain the male trait away from the natural selection optimum (Lande 1981). In finite populations genetic correlations between the male trait and female preference are unlikely to be sufficiently strong, since both characters are affected by drift and the threshold frequency of choosy females needs to be large (Pomiankowski 1988, Nichols and Butlin 1989). As pointed out by Nichols and Butlin, it remains possible even in small populations for a female preference to maintain a character away from the natural selection optimum. But in the absence of genetic correlations, the reason must be some viability advantage for female choosiness as first envisaged by Fisher (1930, 1958).

Pomiankowski et al. (1991) presented a particularly realistic Fisherian type of model, since it allows female choice to be costly and yet aesthetic traits can evolve. The critical assumption was that there is a mutational bias in the genes that make the ornament appear as perfect as possible. No doubt there will be mutations among the genes, and on average it is reasonable to assume that such mutations are biased to impair rather than to improve the attractiveness of the ornament.

Another important recent innovation that may explain the initial origin of aesthetic traits is based on the observation in some fishes and frogs that biases in the sensory apparatus of females may exist before the appearance of signals in the males (Ryan et al. 1990, Basolo 1990). Enquist and Arak (1993) used simple artificial neural networks to explore the possibility that lack of perfection in recognizing conspecific males may produce biases in female preference. The elegant exercise illustrates that the existence of such biases may be sufficient to explain the evolution of aesthetic secondary sexual traits and signals in males.

QUALITY INDICATORS

Models of quality indicators are based on the idea that by choosing ornamented males, females gain by producing fitter offspring in terms of general viability. As mentioned in the introduction of this chapter, the main theoretical argument against this idea is the prediction that without any additional process that creates new genetic variance for general viability, the genetic variance for viability is necessarily diminished over evolutionary time (Fisher 1958, Taylor and Williams 1982, Charlesworth 1987). That such additional factors exist is beyond doubt: recurrent mutation (both deleterious and beneficial), migration, and spatially and temporally varying selection pressures all occur in natural populations (Charlesworth 1987, Pomiankowski 1987b, Iwasa et al. 1991). Whether they are important in sexual selection is still a matter of controversy. It seems that genetic benefits cannot be very large, since traits that are under strong selection do have low heritabilities (Falconer 1981, Gustafsson 1986). However, the benefits only have to exceed the costs of choice in females, and such costs may not be particularly high on leks.

Before presenting in some detail the models for quality indicators, it is necessary to emphasize that males may have these indicators both for genes that influence viability (Iwasa et al. 1991) and for any traits that influence directly the breeding success of the female (Hoelzer 1989, Price et al. 1993, Palokangas et al. 1992). In this section, we will consider the former possibility based on indirect benefits in terms of inheritance of viability-enhancing genes (e.g., Hamilton and Zuk 1982). The direct benefits will be considered later in section 4.2.

Genetic models that invoke a viability advantage assume three loci. Like in genetic models of the pure Fisher process, there is a trait locus and a preference locus and, in addition, a locus for general viability. The first models of this kind assumed a monogamous mating system and equal number of offspring for all males. This was done to study the behavior of preference and male viability and to avoid confusion with the Fisher mating advantage. In the first model neither the preference nor the trait was likely to evolve because of the survival cost to males possessing the trait (Bell 1978). Andersson (1986) made an additional important assumption that changed the evolutionary outcome of the model. In his model, the trait was only expressed in males having the high viability allele on the viability locus and the female preference could spread if the increase in viability attributable to heritable variation in viability was greater than half the reduction in survival caused by the male trait. In Bell's (1978) model this inequality was much greater. The model of Andersson thus showed that preference for a male trait which reduces survival could spread if only expressed in high-quality males, that is, a condition-dependent handicap, even in the absence of a Fisherian mating advantage.

Subsequent models incorporating male viability also allow ornamented males a mating advantage and show that the preference could spread when certain threshold values of the preference were exceeded (Pomiankowski 1987b). A quantitative genetic model of the evolution of female preference, male trait, and viability reached the conclusion that preference for the trait was not possible (Kirkpatrick 1986). However, this model did not allow for heritable variation in male viability. Iwasa and co-workers (1991) showed that when a biased mutation pressure on viability and a direct relationship between the degree of expression of the male character and viability are assumed, costly female mate preferences can evolve.

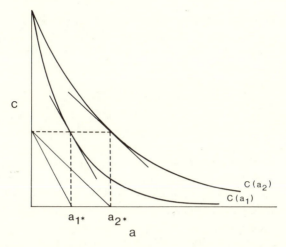

Figure 4.2 Cost of advertisement seen as decreasing survival, *c*, in relation to advertisement level, *a*. If *c(a)* is a negative exponential, at the ESS all phenotypes have the same cost but different levels of optimal advertisement. In this example there are two phenotypes: phenotype 1 and 2, where 2 is better than 1. (Modified after Parker 1983)

A different approach to model sexual selection is to use ESS (evolutionarily stable strategy) theory. Such models could be thought of as a subclass of models dealing with biological signals in general (e.g., between predators and prey, pollinators and flowers, intrasexual aggression; Nur and Hasson 1984, Enquist 1985, Hasson 1989, 1990, Grafen 1990a). The first model of this kind applied to sexual selection is a scramble competition game proposed by Parker (1982, 1983). This game seeks the level of advertisement (e.g., tail length) that is an ESS. Observe that in this model of optimal trait expression in males (and in the following model below), it is irrelevant whether males advertise direct or indirect benefits for females. Furthermore, the evolution of female preference is not what is modeled, but rather how males should signal given that females prefer large traits.

Assume that a male achieves benefits B_i (matings) in proportion to his level of advertisement a_i relative to the mean level of advertisement (mean *a*)

$$B_i = a_i / \text{mean } a.$$

Advertisement also reduces survival c_i which could be generated by energy losses or risk of predation. Assume that $c(a)$ is a negative exponential function. At the ESS, $-c'(a^*) = c(a^*)/a^*$ which means that at the ESS, the gradient $c(a)$ must equal the slope of $c(a)/a$ (fig. 4.2). To explain this more intuitively: given a low level of advertisement, a male can increase his advertisement further at a low cost. His investment can increase further until advertisement reaches a point on $c(a)$ where further benefits by advertising even more are exactly balanced by further costs. This is the ESS advertisement. At the ESS, all phenotypes have the same level of costs, that is, $c(a^*_i) = c(a^*_j) \ldots . \; c(a^*_n)$. Given a large population, the ESS level of a (a^*) is

$$-dc(a^*)/da = c(a^*)/a^*.$$

This means that different phenotypes will have different ESS levels of advertisement. More specifically, poor males will have short tails and good males will have long tails. This model was developed to model passive attraction and thus did not consider the evolution of a female preference. It is assumed that females, by responding to the stimuli, benefit by reducing their search costs. However, the model is in fact more general than proposed by Parker. Given a benefit for females in terms of increased fecundity, the result that males should signal according to quality is a general outcome (see Andersson 1982b, Nur and Hasson 1984, and Grafen 1990a for similar results).

Nur and Hasson's (1984) model of optimal male advertisement is presented graphically in figure 4.3. In this model males gain matings in relation to their level of advertisement x described by the function $F(x)$. Males also pay a cost of x in terms of reduced survival described by $S(x)$. The example in figure 4.3 shows three types of males; S_0, S_1, and S_2, where S_0 is the poorest quality and S_2 are the best males. By multiplying the mating advantage $F(x)$ with the survival cost $S(x)$ for the three classes of males, the fitness $V(x)$ of each class in relation to x was found. From figure 4.3 it is clear that the level of advertisement where $V(x)$ is maximized is different for the three types of males. The optimal level of advertisement x_i and fitness $V(x)$ is higher for the best males than for intermediate males, which in their turn advertise more and gain higher fitness than the poorest males.

The best-known models of quality indicators are based on the verbal arguments of Zahavi (1975, 1977), who suggested that females prefer to mate with a male who possesses traits detrimental to survival because such males signal a better breeding situation to the females (Kodrick-Brown and Brown 1984, Grafen 1990a). This is known as the *handicap principle.* In handicap models, secondary sexual characters evolve not despite the fact that they lower the survival of their bearers, as assumed in models of the Fisher process, but because of it. By being costly the signal is an honest advertisement of a breeding situation. A good breeding situation could simply mean a direct benefit in terms of increased fecundity and/or better quality offspring because of environmental reasons (breeding in better than average territories). However, Zahavi claimed that the handicap principle could work even when the benefits to females are only indirect. By mating with handicapped males, females would raise offspring that inherit the genetic quality of their fathers in terms of overall fitness.

The handicap principle was at first criticized, and most attempts to

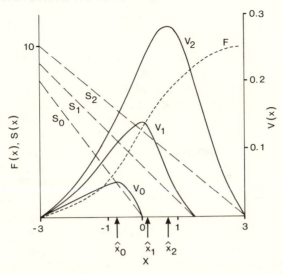

Figure 4.3 Nur and Hasson's multiplicative model of optimal male advertisement. The model assumes three classes ("conditions") of individuals. S describes survival, F the mating advantage, and V fitness (F multiplied by S), and x is the level of advertisement. Each class of males (0, 1, and 2) obeys its own S-curve and generates its own V-curve. x_0, x_1, and x_2 maximize the respective V-values. (After Nur and Hasson 1984)

model the process reached the conclusion that it was an unlikely explanation for the evolution of female preference and exaggerated male traits (Maynard Smith 1976, Bell 1978, Kirkpatrick 1986). However, the rejection of the principle seems to be based on the assumption that all males irrespective of quality invest equally in the handicap, an idea that goes against a basic assumption of the handicap principle.

Other criticisms of the handicap principle are that ESS models have not provided population-genetic models that explicitly show the behavior of alternative alleles and have not modeled the origin and maintenance of female preference but merely taken it for granted (Maynard Smith 1983, Lande 1987). The first point has been addressed in a difficult but important paper by Grafen (1990b), which provides three population-genetic models of the handicap principle: a single locus model where advertisement is expressed in males and preference in females, and a pair of two-locus models. Grafen showed that population-genetic models conform to simpler ESS models.

Concerning the origin of female preference, Grafen (1990b), like other ESS models of the handicap principle, assumed that preference pays because of direct selection through increased number of offspring. If handicaps advertise male viability, it is quite likely that they could also be positively associated with any direct effects (decreased risk of disease transmission, high sperm quality, etc.) on female breeding success. Nevertheless, the origin of female preference is not a problem only confined to handicap models. Remember that Fisher originally envisaged a fecundity advantage to females to get the female choice spiral started, and a fecundity advantage can initiate female choice (Heisler 1984). However, as mentioned earlier, sensory biases may also initiate aesthetic trait evolution.

Another problem for handicap models is to explain the maintenance of the preference. If the increased number of offspring occurs only for purely environmental reasons, there is no problem. The difficulty appears if female fecundity is raised by only genetic factors. As previously stated, quantitative genetics theory predicts low genetic variation for general viability. However, the seriousness of this argument depends on how often in nature there will be only environmental reasons for choosing ornamented males, and if zero heritability of general vigor is ever the case.

To allow females to choose among males, signaling must be honest (Grafen 1990a,b, Zahavi 1991). This is to say that it is not an evolutionary stable strategy for poor males to cheat and signal at high levels. Given honest signaling, rather than assessing a prospective breeding situation directly, which may be impossible or time consuming, females that pay attention to male signals achieve better breeding situations. Better breeding situations are found among the best males, which signal more and have higher viability (Nur and Hasson 1984, Grafen 1990b; but see Parker 1983 for a case where viability is the same). Thus the handicap principle predicts a positive correlation between survival, advertisement level, and mating success.

Are all viability-linked signals in sexual selection handicaps? The answer depends partly on the definition of a signal. Maynard Smith (1958) distinguished among advertisements and "signs." The former are signals that are manipulative, the latter are informative signals. Only advertisements need to be costly to be honest. Hasson (1989, 1990, Michod and Hasson 1990) distinguished between amplifiers and quality indicators. In Hasson's genetic models, traits that amplify previously existing fitness differences can spread without being costly (Hasson 1989). An example of such a character is barring and lining of feathers in avian plumages (Hasson 1991). Such characters are probably expressed at a low cost but would help a female to perceive differences in plumage quality, which is presumed to be a reliable and costly signal. The amplifier by itself is not attractive to females. Amplifiers can spread both with and without a Fisherian mating advantage (Hasson 1990).

Consensus?

The debate in sexual selection in the last decade has largely centered around two so-called opposing schools: a world according to Fisher-Lande, and a world according to Zahavi. This dichotomy is not strict: many models that invoke a viability advantage to a large extent depend on the Fisherian mating advantage, whereas other viability models dismiss the run-away process entirely. Similarly, many Fisher models rest on a viability advantage, at least initially.

Our own view is that the runaway aspect of Fisher's argument is probably overemphasized. However, no one has yet worked out the ultimate reason why females prefer some males over others in any lekking species. Indeed, this question is unsettled with regard to the entire animal kingdom, and the reasons are likely to differ among species. On leks, as in other mating systems, there may be both direct and indirect benefits for females to choose among males (see the next subsection). We believe handicap-type models probably come closer to what is actually happening in leks, even if it is unclear to which degree the choosy females are gaining direct or indirect fitness benefits. However, there are no direct tests where the fitness consequences of female choosiness have been estimated under natural conditions. As argued in the previous chapter, females in some species seem to direct their mate choice at traits that are not arbitrary but costly to produce and often positively associated with male viability.

A reasonable interpretation (see Møller and Pomiankowski 1993) is that costly characters such as continuous display and long tails represent cases of honest signaling and are therefore best thought of as handicaps. Characters that, on the other hand, are cheap, such as feather lining, are probably best thought of as amplifiers or aesthetic characters, possibly explained by Fisherian mating advantage (Alatalo et al. 1988). Both types of traits may be found in males of lekking species, but we believe the former are more influential in mate choice.

DIRECT BENEFITS

Much of the controversy about female choice has been over questions that concern the genetics of the process, in particular whether females can choose males on the basis of heritable viability differences. Indeed, this question is why many lek studies were begun. Lek species have been suggested to be ideal for such studies since it has been assumed that the only possible benefit for females that choose is genetic. However, the foundations of this claim have, surprisingly, not been well studied.

In many studies of species with resource-defense mating systems it seems as if females choose males on the basis of territory characteristics that improve their breeding performance (e.g., Alatalo et al. 1986b, Thompson 1986, Warner 1987, Hastings 1986). Even if such choices could be based on direct assessment of territory quality, results on the pied fly-catcher suggest that females in this species use the quality of male song as a mean of choosing a good territory (Gottlander 1987, Alatalo et al. 1990a, Eriksson 1991). In birds of prey the nesting success of females is dependent on the hunting success of the male that provides the food for the female and

chicks until the late nestling stage. In the kestrel, females choose ornamented, brightly colored males, and these males are most successful in
bringing food to the nest. Thus females increase their offspring production
by choosing ornamented males (Palokangas et al. 1992). A further possible
example where males may signal the possession of good breeding opportunities is in the avoidance of parasitic infection (Zuk et al. 1990a,c). Is it
possible that in lek species males possessing exaggerated traits signal good
breeding situations to females, and therefore female preference could be
explained by immediate benefits?

Two recent reviews have emphasized the possibility of direct benefits of
female choice in lek species and list a number of possible mechanisms
(table 4.1; Reynolds and Gross 1990, Kirkpatrick and Ryan 1991). Among
these possibilities not much is known about reduced search costs, the risk
of injury, sperm quality and quantity, and disease avoidance; but some
evidence suggests that predation risk, hybridization, and pleiotropic effects
on female choice behavior could play a role in some systems.

The reduced-search-cost explanation is similar to the passive-attraction
model of sexual selection (Parker 1983). However, female sampling and
rejection of males within leks argue against this being common (Trail and
Adams 1989, Alatalo et al. 1991, Petrie et al. 1991). For similar reasons, it
is unlikely that hybridization avoidance can explain the maintenance of
female preference in most lekking species. Females, according to this explanation, show preferences because they are selected to reduce the time it
takes to find a conspecific mate and avoid hybrid matings. However, while
such an explanation may have some validity in, for example, anuran choruses or aggregations of insects where several species can breed in the same
pond or restricted area, it is an unlikely phenomenon on vertebrate leks.

Females on leks may prefer certain males because they may provide
more sperm and/or better-quality sperm (Avery 1984, Barnard and Markus
1989). For either reason, females would have an immediate benefit of mating with such males. In many species, testes size and thus sperm supply is
related to age and physical condition (see references in Avery 1984). It is
possible that by mating with males that signal old age and vigor, females
avoid the risk of not being fertilized or receiving sperm of low quality.
Avery also stressed that on leks, matings are often disrupted, and disrupted
matings are likely to have costs in terms of infertility. Perhaps females gain
by choosing dominant and vigorous males that run low risks of disruption
during copulation.

The testes of sharp-tailed grouse at the center of the leks contained more
spermatozoa than the testes of males at the periphery (Nitchuk and Evans
1978). Also, in the same study, it was found that the testes of males on large
leks contained more sperm than testes of males in small leks. Assuming
these differences correspond to male fertility, the results suggest that mat-

Table 4.1

Possible Direct Benefits of Female Mate Choice on Leks and the Corresponding
Male Trait Such Benefit Is Likely to Favor

Female Benefit	Male Trait under Selection
Reduced search costs	Conspicuousness
Reduced predation risk	Traits correlated with defending safe mating territories
Risk of injury	Intramale dominance/honest signals thereof
Sperm quality/quantity	Male fertility/honest signals thereof
Disease avoidance	Disease resistance/honest signals thereof
Reduced risk of hybridization	Species-specific signals
Results of pleiotropic effects on sensory system*	Sensory exploiting traits

*Preference a result of natural selection benefit in contexts other than mate choice.

ing with more competitive males could gain female sharp-tailed grouse an immediate benefit. Similarly, a correlation between male display intensity and sperm quality (a high proportion of the spermatozoa being alive and showing high motility) has been suggested in capercaillie (Mjelstad 1991). Again this result suggests females may achieve fertility gains by discriminating among males, in the latter case by choosing males who show a vigorous display.

Practically nothing is known about sexually transmitted diseases in wild lekking animals. However, populations of commercially kept chickens (*Gallus*), turkeys (*Meleagris*), and geese (*Anser*) provide widespread evidence for existence of pathogenic sexually transmitted diseases (Sheldon, 1993). Pathogenity varies considerably, but goose venereal disease (*Mycoplasmosis cloacale*) can cause up to 10% mortality in affected flocks, together with reduced weight and fertility (Stipkovits et al. 1986, Marius-Jestin et al. 1987). Sexually transmitted diseases include any pathogens that are transmitted during the act of copulation, and thus they also include other diseases than those transmitted in semen. During the copulation, ectoparasites may be transmitted, and in birds with cloacal contact it is possible that gut parasites may be transmitted. Thus the cloaca serving a common function for gamete transfer and excretion may increase the number of pathogens that could be transferred in birds compared to mammals (Sheldon, 1993). The vigorous preening after copulation, which is very obvious, for instance, in the lekking grouse, may in fact serve to remove ectoparasites that have been transferred during copulation (Read 1988).

A recent explanation for female preference is that preferences arise as a correlated response to natural selection (Ryan et al. 1990, Ryan and Rand 1990, Ryan 1991b). This is a phenomenon of pleiotropic gene action. According to this explanation, females have sensory biases due to natural selection. For example, females may respond to moving objects because this gives them an advantage while hunting for prey. Males, on the other hand, are selected to exploit such preexisting biases in the female sensory system when advertising their willingness to mate. Once a female preference is established, this will determine the equilibrium expression of the male trait and can cause the evolution of male traits that reduce male survival (see references in Kirkpatrick and Ryan 1991). Empirical evidence of the sensory-exploitation hypothesis has been claimed in studies of tùngara frogs and swordtail fish (genus *Xiphophorus*) (Ryan et al. 1990, Ryan and Rand 1990, Basolo 1990).

If direct benefits of female preference turn out to be common on leks, the theoretical dispute in sexual selection becomes less important. What is needed at this stage is more empirical data that address the question of direct benefits of female choice behavior. If direct benefits to females in lekking species turn out to be common, female choice on leks may be no paradox after all.

INDIRECT BENEFITS

Evidence for indirect benefits in any lekking species is still meager. This is mainly explained by the fact that such data are extremely difficult to collect. Since lekking females leave the lek and take care of the offspring without male assistance and because lekking species are often precocial, it is difficult to catch a sufficient number of offspring and assign paternity to them. Even if offspring number could be determined, this leaves the researcher with the difficult task of following offspring performance, something that is not easy even in species that, for example, breed in nestboxes.

In black grouse (Alatalo et al. 1991) and peafowl (Petrie et al. 1992) it has been shown that females prefer to mate with mates that are more likely to survive. Such data, even though consistent with an explanation of choice of good genes, are open to other interpretations. For example, males that show high survival could be males that are less likely to transmit diseases to females while mating, and the female preference for such males could thus have to do with a direct benefit. To show that females gain indirect benefits it is necessary to show that the choice of females affects offspring performance while controlling for possible confounding effects such as maternal effects and differential reproduction investment.

This has so far been done in only one species. In peafowl, the offspring of highly ornamented males tend to grow better. This translates into differences in the chance of their subsequent survival under seminatural conditions (Petrie, in press). Thus, at least in this species, female preference is best explained by an indirect good viability genes explanation. It remains to be shown if this result is a general outcome also in other lekking species.

THE NATURE OF FEMALE PREFERENCE IN LEKS

What can be learned from the knowledge accumulated so far of male traits and female preferences on leks? It is perhaps a bit disappointing that no firm tests of existing models have been produced. Part of this difficulty lies in the nature of the models. Predictions and tests proposed so far are often not mutually exclusive, and even if data are compatible with one hypothesis ad hoc reasoning can usually save the alternatives (Kirkpatrick and Ryan 1991, Balmford and Read 1991). Another difficulty lies in the problems of studying recruits with known parentage and their performance in the wild. In practice this requires large research programs where all males and females are caught and marked, reliable methods of determining parentage, and the finding and marking of offspring. If this is accomplished, sufficient sample sizes will be needed to perform the relevant parent-offspring analyses to determine the heritability of the characters under study.

Observe that matings controlled by the researcher are necessary before true heritabilities can be estimated, since otherwise it is possible that any association between female and male qualities may increase parent-offspring resemblance. If in addition there are maternal effects, it will even be impossible to say if there is any heritability at all. In fact, studies of the black grouse have indicated that older and heavier females that are capable of laying larger clutches are more likely to mate with the most attractive males than are females with small clutches (A. Lundberg, J. Höglund, R. Alatalo, and P. Rintamäki, unpubl.). A further problem is that females may alter their reproductive investment in relation to the quality of their mate (Burley 1986, 1988). Recently this possibility has been suggested in experimental studies in the barn swallow (de Lope and Møller 1993) and the peacock (Petrie and Williams 1993). Therefore it would also seem necessary to eliminate the possibility that females are aware of the quality of their mate. Such data are extremely hard to obtain in studies of wild populations.

However, it is a mistake to consider as failures the studies that have been carried out so far on female choice on leks. We know more about male traits and female choice today than ten years ago, and we have gained the knowledge to ask the relevant questions. Scientific research would be easy if we always knew where to look for the answers.

Today we know that females do choose males on leks, at least in some species (chapter 3), but we know little about the costs of choice (see section 4.3). One study suggested that the costs to females were low (Gibson and

Bachman 1992). More studies are needed to confirm if this is a general result. Recent studies suggest that female choice is likely to be based on a number of male criteria (dealt with in detail in chapter 3). Also, male secondary sexual traits often appear condition dependent, which needs to be explained in theoretical modeling. Finally, male success seems to be linked to survival abilities. This means that males with large ornaments do not suffer increased mortality, which supports the findings of theoretical models of condition-dependent ornaments. Fitting these findings into a theoretical framework is a future task for theoreticians, while testing this theory with new empirical findings is the challenge for field workers.

4.3 Costs of Female Choice

When studying the evolution of female preferences it is essential to consider the magnitude of the costs when females are choosy. The idea of the lek paradox (Taylor and Williams 1982) emphasized the problem of understanding female choosiness, when the benefits may be very small indeed. In particular, this is the case if females are choosing for "good genes" for the offspring. Kirkpatrick (1985, 1986) altogether dismissed the possibility that there could be any heritability for fitness, but more realistically the question is how small the heritability is. On the other hand, the early Fisherian models (Lande 1981, Kirkpatrick 1982) assumed that there is no cost at all for females to be choosy, which is impossible in any realistic situation. If cost is included it is impossible for these models to create aesthetic traits, and it is only in the recent model of Pomiankowski et al. (1991) that aesthetic traits evolve.

Against this framework it is necessary to consider the cost for females and relate it to the various types of benefits that females may achieve. In theory there are, in this context, two kinds of costs to lek-breeding females: costs of mating on leks and costs of choosing males within leks. However, in practice separating between the two is often impossible. Not much empirical work has been devoted to study the such costs and the few exceptions have focused on the costs of mating on leks. Gibson and Bachman (1992) identified two costs when comparing a typical series of lek visits with a hypothetical single short visit in sage grouse. The extra travel increased the predicted daily energy expenditure by only about 1%, and the risk of predation by golden eagles was estimated to reduce annual survival by < 0.1%. Furthermore, repeated lek visits did not depress foraging time or conflict with nest defense.

The results may be typical for bird leks, but further studies are needed to confirm this. Even if females typically visit the lek(s) over a period of several days, the visits make up only a small portion of the day. The energetic costs are likely to be small, given that in most lekking birds females use fruits or other plant material as food and have to feed only a part of the day to fulfill their energy demands. Predation costs on females are likely to be small, since even males in only some species (e.g., cock-of-the-rock; Trail 1987) seem to suffer from predation. In the black grouse we have followed about thirty seasonal leks during the main mating week. Goshawks frequently attack the leks, and they direct their attacks at displaying males. However, we have never seen a successful case of predation, even though males have been physically contacted by the hawks on two occasions. Once we found the feathers of a male on the arena verifying a killed male, but females have never been killed. On the contrary, outside the leks twelve out of sixty-three females were killed in May (P. Valkeajärvi and L. Ijäs, pers. comm.), suggesting that females run less risk of predation on the lek than at other times.

In ungulates, harassment by males may even make it beneficial for females to stay at the lek during the rut, and ungulate females do stay on leks relatively long times (Clutton-Brock et al. 1992). In general, if lekking will cause predation risk, that risk should be higher for males. However, costs of choice are not trivial, if we recall that females are quite limited in the time in which fertilization can take place. In capercaillie, copulations take place about four days before the laying of the first egg (information from captive birds, S. Ylisuvanto, pers. comm.). In the black grouse it seems that females are very insistent about getting the mating on a specific morning. This view is corroborated by the fact that if the copulation is interrupted so that the female has to remate, she will typically do so during the same morning or not later than the next morning (R. V. Alatalo, T. Burke, J. Höglund, and A. Lundberg, in prep.).

In animals that are less mobile, the costs of visiting the leks may be higher, and indeed that cost may abolish the possibility for leks to occur in many invertebrates or reptiles. This may be one of the reasons why classical leks are quite uncommon in insects, in spite of the lack of paternal care. Females cannot pay high costs in their mate choice, and instead matings take place at the sites of emergence, feeding, or oviposition. On the other hand, if population density is low and oviposition sites are well scattered, as in parasitic wasps, the use of landmarks as mating arenas may evolve to reduce costs for female mate searching (Thornhill and Alcock 1983).

In summary, it seems that females do not pay particularly high costs for choosing mates. Thus even slight direct or indirect benefits of choosiness may be sufficient to explain female choosiness, and leks may not be paradoxes at all. Since costs of checking further leks are likely to increase linearly or even disproportionally against the number of males compared, while the benefits are likely to increase only asymptotically, we can expect the optimal sampling by females to be rather small.

4.4 Female Mating Strategies

In the following sections we will consider some basic features of female mate choice strategies. In this discussion we will assume that females are searching for and assessing optimal males for whatever benefit. The simplest model of such mate assessment is that each female evaluates several males individually, presumably at some cost that has to be outweighed by the benefits. If this is the case we would expect females to have evolved various kinds of cost-reducing behaviors, which include the use of spatial and temporal cues in mate choice.

Cost Reduction and the Use of Spatial Features

First we need to define what we mean by cost reduction in female mate selection. Imagine a female visiting a lek to mate. Assume that mate selection is not random but based on the assessment of some property of males. In other words, females are ultimately seeking to mate with the kinds of males that maximize their fitness. Females visiting a lek are thus facing an information-sampling problem. We call the mate-sampling behavior pure assessment if a female bases her mate choice on the information provided by the males present, that is, if she visits all or most of the males on the lek and chooses her mate according to some stable rule (Janetos 1980, Real 1990). The cost of pure assessment could be measured as the time and energy devoted to this behavior. If, on the other hand, females reduce the search costs to find the best male by using information from previous visits or by other females' behavior, the female is using a cost-reducing strategy.

An example of a cost-reducing behavior may be to use spatial information of the males on the lek. A common suggestion is that females use the density of the lekking males, which increases toward the center of the lek, to choose the best males (e.g., Kruijt and de Vos 1988, Gosling and Petrie 1990). Superficially such a process seems to require that males to some extent cooperate and that peripheral males act to increase the attractiveness of central ones. However, it may be the best strategy that the peripheral males have and as reviewed in the previous chapter, they may have some possibilities to gain matings by just being as close as possible to the attractive central males (spatial spillover). Thus females could use the rough position on the lek to find the best males if unsuccessful males gather around preferred males, but direct tests for the use of centrality as a spatial shortcut are surprisingly few.

Another possibility of a cost-reducing behavior could be to show site fidelity to the territory where one has mated before (Gibson 1992). Such a "temporal spillover" process is most likely to occur in long-lived species or in species where the best males compete for previously successful territories. In sage grouse it has recently been shown that males prefer to settle in territories that have had a high previous success (Gibson 1992). However, temporal spillover is particularly common in African ungulates (Floody and Arnold 1975, Balmford 1990, Balmford et al. 1992, Deutsch and Nefdt 1992), where successful males have to leave the lek after a few day periods of continuous territory defense.

COPYING

It has been suggested that females of various lek-breeding species copy the mate choice of others (e.g., van Rhijn 1973, Wiley 1973, Lill 1974, Bradbury 1981, Pruett-Jones 1992 for birds; and Clutton-Brock et al. 1989, Balmford 1990, Deutsch and Nefdt 1992, Clutton-Brock and McComb 1993 for mammals). If mate-choice copying occurs, it can be present for two not mutually exclusive reasons (Gibson and Höglund 1992). First, by copying others females may reduce the time they sample prospective mates. This time reduction may be beneficial because mate sampling may be costly. Second, mate-choice copying may have nothing to do with cost reduction, but instead increase the precision of assessment. It is possible that while sampling males, females have to collect and process information about a number of male attributes. Mate choice is thus viewed as a problem of sampling information about different males. While doing this it has been suggested that females should pay attention to what other females are doing since the choice of previously selecting females contains information that others have accepted their males as mates. These males could therefore, for no other reasons, be suspected to meet a given standard (Bikhchandani et al. 1992).

In their studies of sage grouse, Bradbury and co-workers were interested in the unanimity of female choice observed in almost all lek species that had been studied (Bradbury et al. 1985). They asked if female preference for male characteristics alone could account for this variance. In doing this they obtained simulated values of variance in male mating success by assuming values for female discriminating abilities and certain values for male traits and survival. These simulated variances were then compared with observed variances obtained from field data. In almost all comparisons, the observed variance in male mating success was higher than the simulated. Consequently, it was concluded that direct female assessment alone could not explain the large variance in male mating success observed in many field studies, and they suggested that copying, among other possibilities, was one factor that could explain this discrepancy.

Another approach to the question of copying has been to model distributions of male mating success assuming that copying occurs. One such

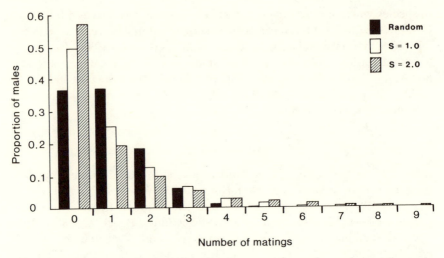

Figure 4.4 Frequency distribution of matings among males under random mating and assuming that females copy the mate choice of others. S measures the strength of copying. (After Wade and Pruett-Jones 1990)

model is the Polya-Eggenburger equation that describes the probability that colored balls are drawn at random from an urn and replaced with several additional balls of the same color (Feller 1968). The color distribution of balls after n selections is given by this model. Copying is analogous to the Polya-Eggenburger equation if the probability that a male is chosen by females depends on the number of times he has been chosen before (Wade and Pruett-Jones 1990). The skew in the distribution of matings is thus enhanced if this is assumed (fig. 4.4).

Despite copying being potentially important, it has been difficult by mere observations to show that copying exists in wild populations. In sage grouse, 64% of all females mating on a given day could potentially have used a copy strategy, since these females were present on the lek when other females mated (Gibson and Bradbury 1986). If copying occurs, it could be argued that the mating skew on given days should increase with the number of females mating. Alternatively, if each female chooses her mate independently of other females present on the lek, the distribution of matings should be unrelated to the number of visiting females. No evidence that the mating skew increased with the number of visiting females was found on a given day. However, this test is indirect, as pointed out by the authors; for example, it was not possible with the given data to test if females could copy the mate choice observed on previous mornings.

Later studies of sage grouse were more successful in showing that some females copied others (Gibson et al. 1991). The proportion of females mating with males already chosen by another increased with the number of females mating on a given day and exceeded the values expected from independent female choice. Also, the standardized variance in male mating success increased as an index of female crowding increased, while the proportion of variance explained by differences in an epigamic trait, inter-pop interval (chapter 3), decreased. These findings support the copy hypothesis. In fallow deer, individual males were more likely to attract new females into their territories at times when many other females were already present (Clutton-Brock et al. 1989). Again, this supports the notion that females are attracted to males that other females have chosen, and it indicates that females may copy the presence of other females in male territories, and not only copulation acts.

Our own studies showed possible copying in the black grouse. We analyzed the sequence of matings on one lek in 1989 (Höglund et al. 1990b) and hypothesized that if females copied one another the probability that a male who haded a mating would mate again in sequence should increase. Conversely, if each choice was independent, the sequence in which a particular male mated could be expected to be random. In a larger data set from several leks and over many years, the sequence of matings was indeed found to deviate from a random expectation (table 4.2). In five of nineteen leks the number of matings in sequence was significantly higher than expected by chance (fig. 4.5c). This does not mean that copying does not occur on the nonsignificant leks, rather that its presence is hard to detect with statistical methods. If the more general hypothesis that matings should be found in sequence more often than expected by chance on all the leks taken together is tested against the nullhypothesis of random sequences, mating sequences are indeed different from a random expectation (Wilcoxon Signed Rank test, $z = 1.99$, $P = .046$).

However, any of the above-mentioned results could also be explained by alternative hypotheses. For example, it is possible that traits such as display rate and fighting success, which may be important in direct assessment, vary with the physical exhaustion of males during the mating season. That the most successful males on the black grouse leks mate less often toward the end of the season indicates that this could be the case (Höglund et al. 1990b).

To study the prevalence of copying in more detail in black grouse, we designed an experiment using stuffed female dummies (Höglund et al., in press). Black grouse cocks readily accept stuffed females and immediately display vigorously toward them and after some minutes of courtship begin to copulate with them. Copulations with dummies can be very long, and we can make males mount the dummies for 20–30 minutes without interruption. Copulations with real females are brief affairs, and mountings last

Table 4.2

Tests for Independence of Successive Matings in Nineteen Lek-Years

Lek-year	No. of Matings	Mating Skew	Observed in Sequence	Expected in Sequence	P
1	24	.55	7	6.17	.41
2	13	.54	5	4.15	.39
3	8	.38	2	1.25	.36
4	27	.26	2	3.41	.88
5	5	.60	2	1.20	.30
6	53	.45	19	12.00	.003
7	20	.40	5	4.60	.49
8	27	.70	13	13.19	.66
9	59	.25	14	8.03	.01
10	10	.55	3	3.24	.71
11	16	.53	4	4.97	.81
12	24	.29	9	5.02	.04
13	32	.27	11	6.34	.03
14	8	.38	2	1.25	.36
15	12	.75	5	6.16	1.00
16	9	.44	1	1.54	.90
17	11	.64	6	4.84	.32
18	7	.43	4	1.76	.04
19	6	.33	1	0.69	.54

SOURCE: After Höglund et al., in press.

NOTES: All lek-years: Chi-squared = 58.22, d.f. = 38, p < 0.025 (Fisher's test). For each lek season the numbers of runs of consecutive matings by the same male are compared to the average number expected if successive females chose independently of one another. Expected values were computed by simulation. Probabilities of observing as many or more runs as were actually observed, given the assumption of independent choice, were obtained from the simulations.

only a couple of seconds. Thus we can mimic a situation in which a male is copulating and has a group of females in his territory and be certain that the males copulating with the dummies are seen by visiting females.

The results of these experiments show that by presenting dummies to males we can increase their rates of visits by real females. The proportion of visits to experimental males is higher when models are present compared to the day before the experiment and after (Friedman Two-Way Analysis of Variance, $\chi^2 = 7.79$, $P = .020$; fig. 4.5a). The difference between before and during the model presentation, as well as during and after, was significant (Wilcoxon Signed Rank Test, $T- = 0$, $P < .02$ and $T- = 1$, $P < .05$, respectively).

In the above experiment, males were allowed to copulate with the females. To see if females were using the information that males were mating

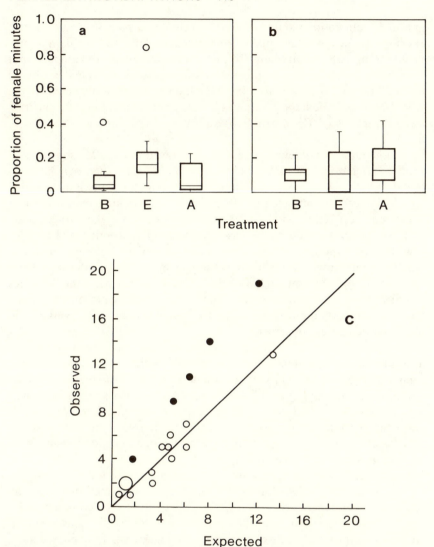

Figure 4.5 (a) and (b) Female responses to the presence of dummies on male territories. The response measure is plotted in relation to day in treatment; *B* is the morning before the models were presented to the focal male, *E* is when models were present, and *A* is the morning after. (a) Female responses when males were able to copulate with the dummies, and (b) when they were not. The box plots give the median, 25 and 75 percent interquartile ranges, and outside values (open circles). (c) The expected number of matings in sequence versus the observed number of matings in sequence. Each point is one lek-year except the large one, which represents values from three lek-years. The filled points are lek-years where the observed number of matings in sequence significantly exceeded the expected number. The line of equal values is indicated. (From Höglund et al., in press.)

or if male solicitation would be enough to elicit a response, we ran a different set of replicates where males were given model females but were prevented from mating with them. To prevent the males from mating, we put the models on sticks 30 cm above the ground, resembling females who sit in small trees or bushes on the leks, something that commonly occurs in the wild. In this treatment the result was less clear-cut and did not demonstrate a significant effect (Friedman Two-Way Analysis of Variance, NS, $N = 7$; fig. 4.5b).

Although the results when males were allowed to copulate showed that visitation rates increased, all but one of the seven experimental males tested failed to gain any copulations by real females. This indicates that females indeed are attracted to other females and mating males, but that being seen copulating with a model is not enough to make females mate with the males. It could be the case that on close inspection real females saw that the experimental male was copulating with a dummy and was repelled by it. However, black grouse females visit the leks during several mornings (Höglund et al. 1990b) and could have mated when the dummies were removed. That visitation rates can be increased by presenting female dummies supports the idea that copying explains some of the variance in male mating success observed on leks.

Several explanations why females should copy have been suggested. This question is dependent on whether the first choices, which by definition cannot be explained by copying, are random or not. If the first choices are indeed random, copying could still explain the high variance in male mating success, but would not favor particular male traits. The present evidence suggests that copying occurs in conjunction with direct assessment (Gibson et al. 1991). Hence copying could be seen as a factor that enhances the mating success of particular males.

Losey et al. (1986) were the first to suggest a reason for why some females should copy others. Assuming that direct assessment and copying entail different benefits but also different costs, they showed in a simulation that copying could yield the same net benefits as direct assessment. Not surprisingly the success of copying strategies was frequency dependent. There must be a sufficient proportion of females who make direct assessments to give copiers the same or larger net benefits. Copying, according to this view, could be seen as a conditional strategy that pays when direct assessment is costly in terms of time and energy and when the majority of females provide copying females something to copy (see also Pruett-Jones 1992). Thus direct assessment may be expected in females that can sustain the costs of choice, whereas copying could be seen as a high risk short-cut strategy used by females who cannot sustain the costs of direct assessment. That young females in both sage grouse (Wiley 1973) and black grouse (Höglund et al. 1990b) mate later than older ones and thus are the potential copiers suggests that conditional factors could be important.

Information concerning the costs of mate search is limited. The only study that has attempted to estimate the energetic costs of mate search in a lekking species found the costs to be negligible (Gibson and Bachman 1992; see section 4.2). However, it is unlikely that females searching for mates face only energetic costs. The time available for mate sampling is not unlimited. After a given time t the benefits of sampling further prospective males must be outweighed by the increasing costs, if nothing else because all biological organisms have limited lifespans. In birds, fecundity commonly drops as the breeding season progresses, probably because of changing environmental conditions. Also, since many lek birds mate only once, there is a limited period of time during which females have to mate or else some or all the eggs in the clutch may be unfertilized. Thus, it is likely that females searching for mates in lekking birds do face costs while sampling. Whether these costs will make some females prone to copy the mate choice of others remains to be shown.

Finally, copying may occur because the mate choice of others contains information about particular males (Bikchandani et al. 1992). Under this model all females assess males and choose them with a certain probability of making an error. By observing the mate choice of others the probability of making an erroneous choice can be reduced. The model shows that after only a few previous matings, females should ignore their own information and choose the male chosen by the others. In this way it is not necessarily the males who meet the female standards who get chosen. Instead, informational cascades can be started by chance, and copying may add noise to the standards by which males are sexually selected. Thus, contrary to the other hypotheses of why copying occurs, which reinforce sexual selection of particular male attributes, an informational cascade may distort sexual selection in males.

MULTIPLE MATING

It appears that in lekking grouse, females copulate only once if they are not subjected to disruption during copulation (Birkhead et al. 1987, Møller and Birkhead 1991). In capercaillie a single ejaculate is enough to fertilize an entire clutch of about nine eggs (Parker et al. 1989). Data on copulation number needed for fertilization are missing for almost all other lekking birds, but the fact that multiple matings are very rare among lekking grouse suggests that a single copulation is also sufficient in other species (Lumsden 1965, Kruijt and Hogan 1967, Robel 1970, Hamerstrom and Hamerstrom 1973, Wiley 1973, Robel and Ballard 1974). In other species

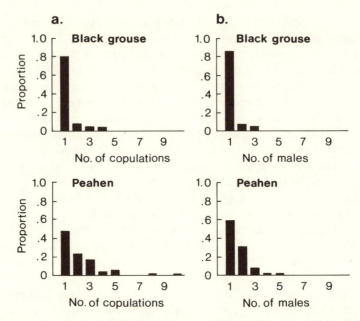

Figure 4.6 (a) The distribution of the total numbers of copulations by individually marked females for a single clutch in black grouse (N = 36; R. V. Alatalo, T. Burke, J. Höglund, and A. Lundberg, in prep.) and in peahen (N = 87; Petrie et al. 1992). (b) The distribution of the total number of males with whom these females mated during the same period.

females may mate with the same male several times during a single or several visits (e.g., great snipe: Höglund and Robertson 1990b; golden-headed manakins: Lill 1976).

Black grouse is one of the species where females typically (86%) mate with only one male for each clutch (fig 4.6). In most cases (81%) females copulated only on a single occasion (see also Kruijt and de Vos 1988). Typically females mate several times only if the initial copulation was disturbed. Similar results with a high proportion of females (> 70%) mating with only one male have been found in the cock-of-the-rock (Trail and Adams 1989), sage grouse (Gibson and Bradbury 1986), sharp-tailed grouse (Gratson et al. 1991), Lawes' parotia (Pruett-Jones and Pruett-Jones 1991), white-bearded manakin (Lill 1974), and golden-headed manakin (Lill 1976). In black grouse, DNA-fingerprinting analyses of eleven broods have confirmed that there is no multiple paternity within the broods, and that the males observed to copulate with the female are the real fathers (R. V. Alatalo, T. Burke, J. Höglund, and A. Lundberg, in prep.).

At least two lekking bird species stand out in respect to copulation frequency and the number of males with which copulations are performed. In peafowl and in ruffs, females may mate several times not only with the same male but also with several other males (ruffs: Hogan-Warburg 1966, van Rhijn 1973, 1991; peahen: Petrie et al. 1992). In peafowl, females show large variation in mating frequency; many females mate with one male (59%) whereas others mate many times with up to five males (fig. 4.6).

Very little is known about why this is so for these two species. It may simply be a fact of insufficient abilities to store sperm in peahens and reeves, but other explanations are also possible. In ruffs we suggested that multiple visits and copulations may be a strategy to probe male quality on the leks (chapter 4). However, the male and female visitation patterns to peafowl leks make this explanation less compelling for this species. Instead it is possible that multiple mating in peafowl could be understood as a mean of reproductive competition among females for the best males (Petrie et al. 1992.). Females may have chosen to mate with an unpreferred male during an early visit because the preferred male was occupied courting other females. If such a female has time and energy to visit the lek again, she may seek another copulation with a higher-ranking male. The first copulation may then be seen as a bet-hedging strategy to countervail the risk of being unfertilized, whereas future copulations are strategies to improve their breeding situation. This hypothesis rests on the assumption that there is last-male sperm precedence, something that has been found in other birds (Birkhead et al. 1987, Birkhead 1988).

Petrie et al. (1992) suggest that by mating with the same male, females prevent other females from mating with preferred males and thus improve their relative success compared to others. Such a behavior is equivalent to spite (Hamilton 1970). Spiteful behavior can only be an ESS in small and structured populations. It remains to be shown if peafowl populations generally meet these assumptions.

Regardless of the explanation of multiple matings in any of the species, the fact that multiple mating does occur suggests that copulation frequency is a trait that can be subjected to selection. Therefore such data strongly suggest that the mating behavior of females can be altered and should not be seen as a black box in models of sexual selection and the evolution of mating systems. In table 4.3 we have listed the possible benefits and costs associated with multiple mating in lekking birds. In fact, it is somewhat surprising that females rely only on a single mating in grouse species, since there could be many advantages of repeated copulations with other males. The pattern of mating singly suggests that costs of multiple mating are relatively high. Among the potential costs, the costs of extra time needed for a further copulation should not be particularly high, since females have assessed the males already for the first copulation, and the same information could be used to pick up another male for a copulation that takes only a very brief time in birds. Therefore sexually transmitted diseases appear as

Table 4.3

Costs and Benefits of Mating with Multiple Males in Lekking Birds

BENEFITS
1. Fertility insurance against poor-quality sperm.
2. Fertility insurance if sperm transfer incomplete because of disturbance.
3. Genetic diversity.
4. Sampling for genetic quality: good genes or sexy sons.
5. Sperm competition.
6. Reduced sperm supplies for competing females.
7. Guard against genetic defects arising from prolonged sperm storage.
8. Excessive costs of avoiding male harassment.

COSTS
1. Disease or parasite transmission.
2. Increased search costs (energy, predation).

SOURCE: Partly modified from Birkhead and Møller 1992.

a more likely candidate for the costs, and no doubt the question of multiple versus single mating deserves more attention in future studies of lekking both in birds and in other groups not considered in this section.

4.5 Summary

The benefits for females to visit the lek and to be choosy among males can be divided into direct and indirect selection. The terminology between direct and indirect effects has been somewhat ambiguous, but with *direct benefits* we refer to any factors that enhance the lifetime offspring number produced by a choosy female compared with a hypothetical nonchoosy female. *Indirect benefits* refer to the fitness differences between the offspring of choosy and nonchoosy females.

The latter possibility has been modeled in the Fisherian approach, where it is only the attractiveness of male offspring and choosiness of female offspring that matter. Alternatively, handicap or "good gene" models assume that the offspring of both sexes in the choosy females inherit viability-enhancing genes. While both kinds of models can create costly female preferences and ornament evolution, there is not much direct information to test between the models.

The obvious difficulty is in estimating the fitness components among the offspring in a satisfactory way under reasonably natural conditions. We feel that the runaway aspect of Fisher's argument is probably overemphasized since females generally seem to direct their mate choice at traits that are not simply arbitrary. However, the directed mate choice on male vigor may also be related to direct benefits for choosy females. These possibilities include reduced search costs, the risk of injury, sperm quality and quantity, disease avoidance, predator avoidance, avoidance of hybridization, and pleiotropic effects on female choice behavior. In practice, it may be that many types of benefits contribute to female choosiness. If the benefits to selective females are direct, there is no paradox of the lek. Future field studies should address this possibility.

In this chapter we have also discussed some aspects of female mating strategies, in particular the possibility that females have evolved various kinds of cost-reducing behaviors while assessing prospective mates. Recently, female copying has been raised as an important factor to explain the surprisingly high unanimity in mate choice.

5

Black Grouse:
A Case Study

5.1 Introduction

As reviewed above, detailed studies of leks have indicated that morphological, spatial, and behavioral traits can be related to male mating success. In some cases these relationships have been confirmed experimentally, indicating that sexual selection is operating directly on the trait in question. In reality, many male traits are likely to be targets of sexual selection and many others are influenced because of their correlation with the selected traits. In this chapter we will summarize the information on sexual selection in male black grouse, a lekking system that has been studied extensively by us (together with Arne Lundberg and Pekka Rintamäki) over seven seasons from 1987 to 1993 (Alatalo et al. 1991, 1992, Höglund et al. 1990b, 1992b).

5.2 Methods of Observation

We have followed annually five to nine leks and made observations of male territorial relationships and display activity. Many of the males have been caught for measurement and color banding. Our main study sites are in central Finland, and here we include data collected until 1992. We have also obtained information from two leks in the province of Uppland in central Sweden; one lek was studied during 1991–1993 and the other in 1992 and 1993. Apart from seeking correlates of male mating success, we have also performed several experiments to test causality.

We caught the males during winter or early April using walk-in-traps, cannon nets, or spring nets. The leks were observed every day during peak female visitation, which occurs annually at the end of April or early May. Most females at any given site mate within 7–10 days in any one year (Alatalo et al. 1992). In both study areas, leks were observed from blinds erected close to the lek sites or from cars or buildings. Males were individually marked with an aluminum ring plus three colored rings on the legs for identification at a distance. On all leks some males remained uncaught, and such males were identified by the individualistic black spot patterns on the white undertail coverts.

Copulation success of the males was scored as the number of females with which a male was observed copulating, and this was recorded on an ad libitum basis. Copulations are conspicuous because the male mounts the female and flutters his wings while copulating. This makes it unlikely that we have missed many copulations on the observed leks. Most females (80%) copulate only once, and there is a good correspondence between observed copulations and true parentage, as revaled by DNA fingerprinting (Alatalo et al. 1992; R. V. Alatalo, T. Burke, J. Höglund, and A. Lundberg, unpubl. data). The number of copulations is therefore a good estimate of seasonal reproductive success in males.

All spatial variables were obtained using a scan sampling scheme where observations were recorded approximately every 15–30 min. Male position was plotted onto maps using a 10 × 10 m grid system on each lek. From these data we calculated two variables: territory size and position. For each observation, the position of all males present was mapped to the closest 1 m on the two-dimensional grid. We estimated territory size for each male as the area in which males spent 90% of their time. Males were assigned a position status as "central" if more than 65% of the perimeter of their territories was surrounded by neighboring males, or "edge" if less than 65% of the perimeter neighbored other males. We chose 65% as the cutoff level because this divided the data in about two equally sized groups.

When a male was sampled, in addition to deciding his position on the lek, we took notes of male behavior. For logistic reasons behavioral data were gathered in the absence of females on most of the leks. We sampled male behavior in the presence of females on only two leks in 1993. Thus the information on male behavior in the presence of females is limited to a few males.

We used a similar behavioral sampling scheme both in the absence and in the presence of females. The exceptions were that male behavior in the presence of females on the lek was recorded only for the male visited by females *and* sampled once each minute. Furthermore, only ethograms were recorded in this situation. The variables—lyre position, eyecomb, fighting level, and attendance (definded below)—were thus recorded in the absence of females only. It should be noted that attendance is probably close to 100% when there are females on the lek. When females are not on the lek some males leave and probably forage or rest nearby.

We noted whether the lyre (the elongated tail feathers and undertail coverts) was kept upright, semi-upright, or down, assigning these positions values of 1, 2, and 3, respectively. Lyre position is the average of all the observations for each male. Similarly, we noted the level at which the red eyecombs were swollen on a five-point scale (from completely folded to wholly erect). These eyecombs consist of modified skin, and their size varies consistently both between and among individuals depending on the activity of the bird (J. Höglund, unpubl. data). Inactive birds typically have folded combs and displaying birds have swollen ones. If the focal male was involved in a fight, we scored the level of aggression on a three-point scale (threats = 1, instant physical fights = 2, and escalated physical fights = 3) and expressed fighting level as the average of all observations for each male. Attendance is the number of scan samples present for each male divided by the score for the male with the highest number on any given lek during the entire season in the absence of females.

We also noted the activity of each male according to the ethogram developed by Koivisto (1965). Activities were as follows: rookoo, hissing, and flutter (advertisement); cackle, fighting, including sparring and threats (fighting); copulation, circling, solicitation (mating); and walking, preening, feeding, sitting (other). All behaviors were expressed as the proportion of these activities for each male. However, in many of the subsequent analyses, we have grouped behaviors into the four main categories: mating, advertisement, fighting, and other.

For each captured male, sternum length (the keel of the breastbone from the anterior to the posterior tip) was measured to the nearest millimeter, and tarsus length (the distance from the heel to the bending points of the toes including the tarso-metatarsal bone) was measured to the nearest one-tenth millimeter with a digital caliper. Wing length was measured with a ruler from the carpal joint to the tip of the longest primary (to the closest millimeter). Lyre was measured as the length of the longest outer tail feather from the base to the tip after flattening it on the ruler. Body mass was measured with a Pesola spring balance to the closest 5 g. For all bilateral measurements we used the measurement on the right side of the body until 1991, when we started measuring both sides of the body to obtain information on fluctuating asymmetry. In all analyses apart from asymmetry, measurements from the right side of the body were used.

5.3 Correlates of Mating Success

Looking at morphological and spatial variables, univariate analyses of males older than one year in age reveal that three traits out of nine were significantly related to male mating success (table 5.1; R. V. Alatalo et al., in prep.). Among the morphological traits only lyre length was slightly positively correlated with mating success (fig. 5.1a), while the other three size traits showed correlations close to zero. In no case were there any significant trends. In this species males are about 30% heavier than females, although sexual selection for increased male size is presently weak or absent. In fact, since size may reflect phenotypic condition as a consequence of growth conditions (Alatalo et al. 1990b), it is not necessary that

Table 5.1

Pearson Product Moment Correlations (r) of Male Mating Success
with Morphological and Spatial Variables in the Black Grouse

	Correlation		
	r	n	P
MORPHOLOGY			
Body mass	.07	133	1.00
Tarsus length	.03	138	1.00
Sternum length	−.04	138	1.00
Wing length	−.02	137	1.00
Lyre length	.11	128	1.00
SPATIAL VARIABLES			
Distance to lek center	−.25	193	.021
No. of neighboring males	.41	186	0
Proportion of territory encircled	.35	186	0
Territory size	−.06	158	1.00

NOTES: Male mating success = total seasonal number of matings ln +1 transformed. Only males over one year old included. P indicates two-tailed Bon-
ferroni adjusted significances.

genes for larger size are under directional selection. In black grouse,
growth conditions clearly influence the size of males since different co-
horts may vary considerably in sternum and tarsus length (R. V. Alatalo,
unpubl. data). We also checked for the possibility of stabilizing selection;
in fact, for tarsus length there was a significant correlation for deviations
from the mean (fig. 5.1b, $r_s = -.20$, $n = 135$, $P < .05$). Males with a tarsus
length close to the population mean were somewhat more successful in
mating than were the males with shorter or longer tarsi.

In combination, the five morphological traits explain only 2.1% of the
variance in male mating success ($ln + 1 -$ transformed no. of copulations,
$F = 1.51$, $N = 121$, $P > .10$). Somewhat surprisingly, the length of the lyre
was not correlated with male mating success (fig. 5.1c) even though this
character looks like an ornament that varies in size. Among the size-related
traits, lyre length is the most sexually dimorphic in the black grouse. The
outer tail feathers of older males are 79% longer ($x = 22.4$ cm, SD = 1.1, N
= 170) compared to females ($x = 12.5$ cm, SD = 0.5, $n = 88$). We have also
checked the possibility that asymmetry in the tail influences male mating
success, but there was no significant correlation ($r_s = 0.14$, $n = 47$, $P > .10$).
Sexual dimorphism is also present in coloration, males being conspicuously black with a white undertail. The red comb of males is another male
ornament, but it is difficult to measure on birds in the hand and is, as mentioned above, seasonally highly variable in expression.

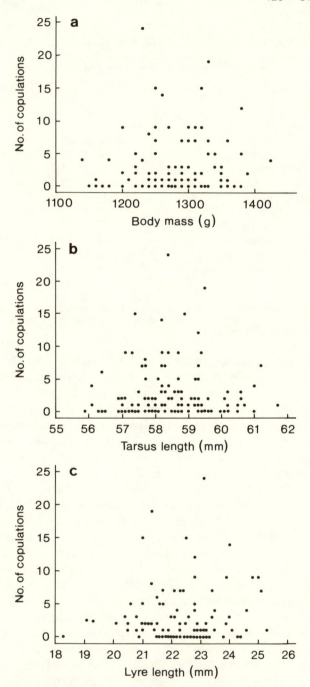

Figure 5.1 Mating success of black grouse males in relation to (a) body mass, (b) tarsus length, and (c) lyre length (for statistics see table 3.3).

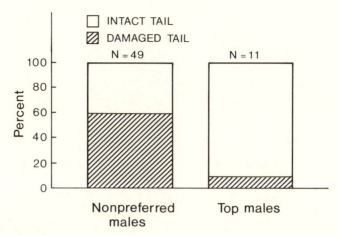

Figure 5.2 The proportion of territorial black grouse males that have observable damage in the tail in nonpreferred males (0–3 copulations) and preferred males (> 3 copulations).

Lyre length is not significantly related to male mating success even in our reasonably large data set, suggesting that ornament size is only weakly associated with male mating success. However, males with damage to the tail, either in the black lyre-shaped tail feathers or the conspicuously white undertail, are unlikely to receive any copulations (fig. 5.2). Thus male ornaments seem to influence mating success through their condition rather than their size.

Among spatial features, three correlated measures of central territorial position (distance to the lek center, number of neighbors, proportion of the territory encircled by others) were all highly correlated with male mating success. Territory sizes are larger farther away from the lek center ($r = 0.46$, $N = 163$, $P < .001$), presumably because of the reduced pressure from fewer neighbors, but territory size alone was not correlated with mating success. The aspects of centrality can be combined into a single parameter using the first axis of a principal component analysis of the three significant variables. When male mating success is analyzed against this measure and territory size, both factors are significant (table 5.2). Male mating success is higher the more central the territory is, the more neighbors surround the male, and the more the territory is encircled by others, and, while controlling statistically for these factors, the larger the territory is.

We should recall that these correlations, in particular with respect to centrality and increased level of male clumping, may represent consequences of female preference if males cluster around the most attractive males. However, if one is to predict the mating success of black grouse males, the best predictions can be achieved by checking the relative positions of male territories and the size of the territory. This can be done even before the females start to visit the lek.

Table 5.2

Multiple Regression Model of Male Mating Success

	Coefficient (± S.E.)	t	P
PC1	3.335 ± 0.524	6.36	0.000
Territory size	0.042 ± 0.009	4.54	0.000
n	163		
adjusted r^2	21.5%		

NOTES: Male mating success is $ln + 1$ transformed. PC1 is the first principal component of proportion of territory encircled, number of neighbors, and distance to the center of the lek. Territory size is ln tranformed.

We analyzed the importance of behavioral variables both in the absence and presence of females using logistic regression (McCullagh and Nelder 1989, Hosmer and Lemeshow 1989). To do this we categorized males as successful (≥ 1 copulation) or unsuccessful (no copulations). Logistic regression models belong to a category of generalized linear models (McCullagh and Nelder 1989) that fit a logistic curve through binomially distributed data in the form

$$p(y = 1) = 1/1 + e^{-Z},$$

where $p(y = 1)$ is the probability of an individual male being successful and Z is the linear combination

$$Z = B0 + B1X1 + B2X2 + \ldots + BPXP.$$

Z is analogous to a standard regression equation, that is, to the sum of a constant and the effects of independent factors, $X1, \times 2, \ldots P$ (and their interactions). Below, different models were evaluated on the basis of the deviance (D), which is proportional to the log of the ratio of the likelihood achievable for the full model and that achieved by the model under consideration:

$$D = \log(L\text{full model}/L\text{model under investigation}).$$

A full model is an exact fit and contains as many parameters as observations. D is thus zero. The null model is the simplest version and contains only the constant ($B0$). If two models are compared, the difference in deviance (G^2) is approximately chi squared distributed under the hypothesis that the simplest model is true. Likelihood-ratio chi square tests may therefore be used to test the significance of improvement if a new factor is added or removed. Here we have used SPSS-X to get the maximum likelihood estimates in the logistic regression models. To choose among models we have used a stepwise option with backward elimination.

Table 5.3

Logistic Regression Models for the Probability of Success
in the Male Black Grouse, Conditional on Male Behavior
in the Absence of Females

Model		D	df	G^2	v	P
Full model		0	0			
Model 1	All variables	23.7	23	6.7	7	NS
Model 2	Other+Advertisement	25.0	28	5.5	2	.06
Model 3	Advertisement	26.3	29	4.1	1	.04
Null model	Constant only	30.5	30			

Variables included in model 1 were lyre position, fighting intensity, eyecomb, other, fighting, advertisement, and attendance. Model 3 included advertisement and the logit is given by $Z = 2.72 - 8.20*$advertisement. D = deviance, df = degrees of freedom, G^2 = difference in deviance from the null model, v = difference in degrees of freedom, P = probability that coefficients for all parameter estimates in the model, except the constant, are 0. NS = not significant.

We ran the first model from behavioral data gathered in the absence of females and entered the following variables: lyre position, fighting intensity, eyecomb, other, fighting, advertisement, and attendance. A model including other and advertisement approached significance and was able correctly to predict mating status in 81% of all cases. A model including advertisement only correctly classified 84% of the cases, and the null hypothesis that the parameter estimate is zero could be rejected (table 5.3). Note that the parameter estimate is negative, thus males that mate are signified by low levels of advertisement (and possibly also low levels of inactivity) (fig. 5.3). Biologically this result could be understood if we recall that the variables other, fighting, and inactivity exclude one another. Mated males were signified by low levels of advertisement. This is probably because unsuccessful males that also occupy peripheral territories spend their time, in the absence of females, unharassed by others, while successful males spend their time fighting and interacting with others.

For the data gathered in the presence of females we entered the following variables: rookoo (i.e., an advertisement behavior where the male stands still while displaying), circling (which is a behavior similar to the previous one except that the male is moving toward and around the female), and fighting. These variables were entered since they, together with copulations, represent the only variables that males perform in the presence of females (fig. 5.4). The model where rookoo was excluded approached significance, while the model including circling only, yielded a significant positive parameter estimate (table 5.4). Successful males were thus signified by high levels of circling in the presence of females (and possibly high levels of fighting). In conclusion, it seems as if males that behave actively,

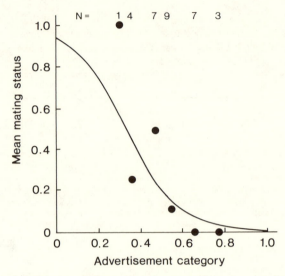

Figure 5.3 Mean mating status (0 or 1) versus mean advertisement for six groups of advertisement categories in black grouse. The line shows the logistic curve given by the logit $Z = 2.72 - 8.20*$advertisement. N gives the number of observations in each group.

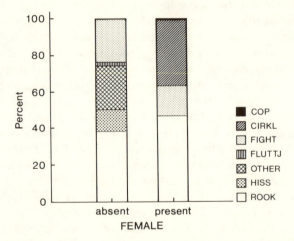

Figure 5.4 Male behavior in the presence and absence of females. Excluding circling and copulations, which can only be performed in the presence of females, and pooling, hiss and flutter jumps reveal a significant difference ($\chi^2 = 230.2$, $df = 3$, $P < .001$).

that is, that move toward and around the female while displaying, are the successful ones, while the males that display standing still were discriminated against. However, it is at this stage impossible to tell if this is a result of females prompting some males to be more active by not leaving their territories, or that active males cause females to stay with them.

Table 5.4

Logistic Regression Models for the Probability of Success in the Male Black Grouse, Conditional on Male Behavior in the Presence of Females

Model		D	df	G^2	v	P
Full model		0	0			
Model 1	All variables	13.7	14	5.4	3	NS
Model 2	Fight + Circling	13.9	15	5.2	2	.07
Model 3	Circling	14.8	16	4.3	1	.04
Null model	Constant only	19.1	17			

Variables included in model 1 were: rookoo, circling and fighting. Model 3 included circling and the logit is given by $Z = -2.35 + 2.97*\text{circling}$. D = deviance, df = degrees of freedom, G^2 = difference in deviance from the null model, v = difference in degrees of freedom, P = probability that coefficients for all parameter estimates in the model, except the constant, are 0. NS = not significant.

5.4 Experiments

While tail length is not correlated with male mating success, there was a correlation with the condition of the tail (fig 5.2). Males may drop some of the tail feathers during predator attacks, presumably to distract the predator. Males may also attempt to tear feathers from the white undertail of their opponents during fights. If a male retreats and starts to run away from the winning male, the winner will attempt to follow and grasp the tail. Similarly, males can grasp other parts of the body, for example, the red comb and the neck feathers, during intensive fights.

We experimentally manipulated the white undertail to test if tail condition directly influences male mating success or if it is just a correlate of other factors influencing attractivity to females (Höglund et al., 1994). On six leks we randomly selected 1–3 males for treatment and cut the major part of the white undertails. We caught the same number of males as we had for controls. Our treatment was very drastic on purpose because very few males have such damaged tails naturally (fig. 5.5). In spite of this, the effects on mating success were marginal (fig. 5.6). In fact, among the central and most successful males the treatment had no effect, and it was only among males at the edge of the leks that treated males did worse than their controls. There is good reason to believe that this effect was due to female preference for intact males, since we did not find any effects of the treatment on male display activity or effects on any territorial features. As a whole, morphological ornaments seem to be of only secondary importance for mating success in male black grouse.

A highly essential experiment for the future is the manipulation of spatial features. As shown above, correlative data indicate that males with

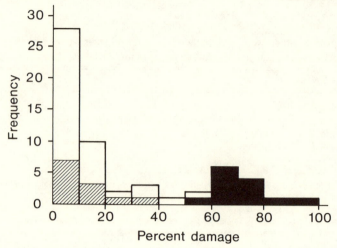

Figure 5.5 Frequency distribution (stacked bars) of damage to the lyre ornament (in 10% intervals). White bars show natural damage (*N* = 35), shaded bars show natural damage to the control group in the experiment (*N* = 12), black bars indicate the damage produced in treated birds (*N* = 12). (After Höglund et al. 1994)

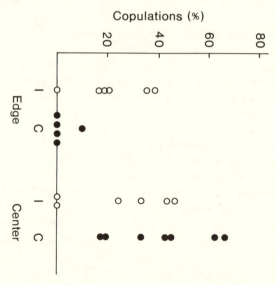

Figure 5.6 The percentage of copulations of central and edge males in relation to treatment and position on the lek (central vs. peripheral). *I* are control males and *C* are males in which a recession was cut into the white undertail coverts of the lyre ornament. (After Höglund et al. 1994)

BLACK GROUSE · 133

relatively large territories in the central area of the lek are the most successful in mating.

Kruijt et al. (1972) performed an experiment that suggests that females
may use male clustering as a criterion in their choice among males. On a
small arena in their Dutch study site they placed two tight clusters of
mounted males, one cluster having six males and the other three. The sizes
of the clusters in the two locations, some 50 m apart, were alternated from
day to day, and thus the experiment was repeated in a rather effective way.
Over twenty-one mornings of exposure, nineteen responses by females
could be recorded when no females or real males were close to the males.
Females approached the large group in sixteen of these cases (binomial
test, $P < .05$). Female black grouse prefer to mate on larger leks (Alatalo et
al. 1992), and likewise they might prefer tighter male aggregations within
leks. However, it is likely that much of the correlation between centrality
and mating success arises merely as a consequence rather than as a cause of
female choice.

There is a common view that on leks females are choosing the most
vigorous and dominant males. This impression comes easily to anyone
watching a lek where males display intensively and where frequently they
also fight with each other. However, not many studies have attempted directly to test for the association between male dominance and mating success (but see Gibson and Bradbury 1987 for data in sage grouse). Testing
the relationship between male dominance in male-male interactions and
mating success was our major goal when we started our studies of black
grouse. The problem is that in a territorial system, each male is dominant
over other males within their territories.

We used two approaches to overcome site-dependent dominance. First,
we studied the dominance relationships of yearling males in a winter flock
just before their first mating season (Alatalo et al. 1991). These males have
not yet established their territories on any lek so they have no association
to any specific site. This study was done within a flock of males that regularly visited a feeding site during the winter. It is a common habit of hunters
in Finland to feed black grouse with oats during midwinter. On the feeding
site we could capture the males and color band them during December and
January. Later in February and March we confined the access of food to a
limited number of spots so that we were able to see interactions between
males in terms of displacement or avoidance. For each pair of males observed to interact we determined the more dominant male. A dominance
score was calculated as an estimate of the proportion of other yearling
males dominated by each male. During the mating season in late April and
early May we followed all the leks in the neighborhood. Four of these
yearling males succeeded in attaining a copulation on leks, and they were
males that had significantly higher dominance scores than males with no

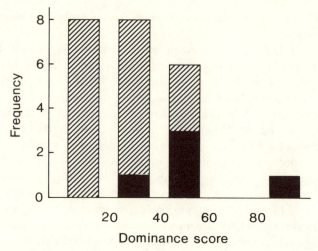

Figure 5.7 The relative dominance (percentage of males dominated by the male in question) in winter flocks for yearling black grouse males that achieved some copulations later in the spring (black), and males that did not achieve copulations (hatched).

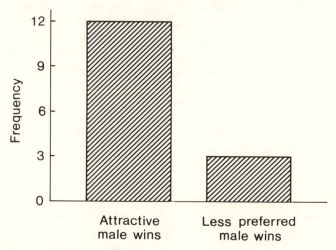

Figure 5.8 The winners in the experiments where a female dummy was placed on the boundary of two territorial black grouse males.

copulations (fig. 5.7). Thus it seems that male mating success is associated with general dominance in male-male interactions.

In another experiment, among older territorial males, we used female dummies to measure dominance between neighboring males. Males are attracted to taxidermic female dummies and frequently start to copulate

with them. In a natural situation territorial males can control access to females inside territories but they can copulate only if females solicit by
lowering their body. Territorial males on larger leks on bogs have very
fixed territory boundaries where males frequently meet for interactions. We
placed a dummy on such a fixed site of territorial interaction where males
typically met for a fight that normally did not show any signs of one male
winning over the other. The idea with the dummy female was to increase
the benefit of the fight for males, and indeed in most cases males started to
fight vigorously over the female. Finally, one male was able to control the
female and he usually started to copulate with the dummy. In twelve of
fifteen experiments the winning male was the one that was more successful
in obtaining mates in a natural situation (fig. 5.8).

On small leks, and in particular on leks on ice-covered lakes without
land marks for territory recognition, males may even intrude onto the territories of their neighbors and thus gain access to females. However, in six
control experiments on the larger bog leks where the dummy was placed
inside the territory of a less successful male, the territorial male was always
able to control access to the female. Thus it seems that females are effec-
tively choosing the most dominant males, even if this experiment did not
directly measure which criteria females were using. Most likely, female
choice is based mainly on behavioral dominance of males in terms of display and fighting ability during the female visits.

5.5 Synthesis

In conclusion, it seems that males that are most vigorous in male-male
interactions and courtship are chosen by females, but that additional factors
such as ornament condition, centrality, and size of territories may be of
some importance. It seems that behavioral traits associated with male dominance are important cues in female choice. Territorial features in terms of
having relatively large central territories may also be relevant, but experiments are badly needed to test this possibility. Morphological ornaments
are probably not of primary importance, but as a whole, females use several
criteria of choice, and thus many male traits have evolved through sexual
selection.

Not much is known about the benefits for females of choosing males that
actively display, are successful in male-male territorial combats, or possess
extravagant ornaments. However, in the black grouse we have shown that
attractive males are most likely to survive a period of six months after each
lekking season (fig. 5.9). The study was done over a period of four years in
an area with no hunting where the main predator on grouse is the goshawk.
The strong association between female preference and male viability sup-

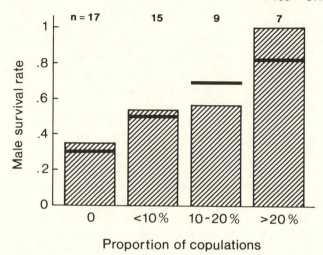

Figure 5.9 Proportion of black grouse males surviving until the next autumn versus the proportion of copulations attained within the lek. The horizontal lines indicate the predictions from a logistic regression model. (From Alatalo et al. 1991)

ports the common view that on leks females choose the most viable males. Similar results have been attained recently in the peacock (Petrie 1992), and in some non-lekking birds male ornaments are also positively associated with male survival (Møller 1991, Göransson et al. 1990, G. Hill 1991). These results support the basic assumptions of the good gene models (section 4.2), but they are also in agreement with studies showing that females choose males for direct benefits.

6 Comparative Studies

6.1 Introduction

The alternative to direct observation and experimentation to obtain evidence of sexual selection on leks is by means of comparative methods (Harvey and Bradbury 1991). Many workers have postulated a relationship between lekking and sexual dimorphism (e.g., Darwin 1871, Lack 1968, Payne 1984). There are, however, few rigorous tests of this relationship. A few attempts have nevertheless been made in testing whether lekking birds in general show sexual dimorphism in size and plumage characters and have reached different conclusions. In this chapter we review such studies.

Postulating a relationship between lekking and sexual dimorphism assumes that lekking is a common selective regime; this is to say that sexual selection on leks will always favor the same kinds of males regardless of species. As we saw in chapter 3, this not the case at all; different kinds of characters are favored in different kinds of species. The multitude of traits favored on present-day leks thus suggest that selective regimes have been different in different species also in the past. With regards to size we have found only three studies (see table 3.1) in which a selective advantage for large males has been established. Inferring past selection regimes from present-day studies is impossible. It may be that larger males have been favored in the past but not in the present. However, we find this possibility unlikely.

There are reasons for sexual dimorphism other than sexual selection. The sexes can be different due to ecological niche separation (Selander 1972). Another reason is that natural selection for a character in females is stronger than sexual selection for the same trait in males. Since fecundity often is strongly correlated with female size, females may be larger in many taxa despite sexual selection for larger males (Ralls 1976, Jehl and Murray 1986, Arak 1988b, Höglund and Säterberg 1989, Höglund et al. 1990c). Such a species would count as monomorphic or even as reversed-size dimorphic in comparative analyses.

Dimorphism, like any other property of species, may show similarity across species due to historical constraints (Pagel and Harvey 1988). In any comparison, a pair of taxa may be similar because they face similar selection or because they share a common ancestor. Thus in comparative studies

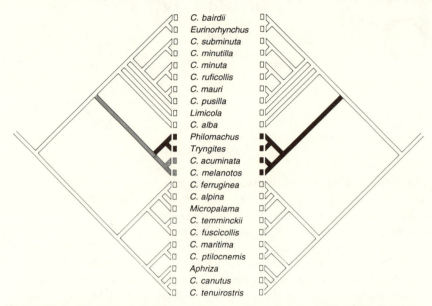

Figure 6.1 A phylogeny of Calidrine waders (modified after van Rhijn 1991). (*Left*) The reconstruction of mating system (white: monogamy; stippled: polygamy; black: lek; and hatched: reconstruction equivocal). (*Right*) Male-biased sexual size dimorphism (black). In this particular example it is equally parsimonious to assume that lekking was derived from a polygynous ancestor or that polygyny evolved from lekking (indicated by the hatched line). However, lekking is a much rarer mating system than resource-based polygyny in the class Aves. Therefore, it is more likely that lekking in *Tryngites* and *Philomachus* once arose in a polygynous ancestor and that polygyny has been retained in *C. acuminata* and *C. melanotos* than to assume the reverse. The transition to male-biased size dimorphism (*left*) coincides with the transition from monogamy to polygyny/lekking (*right*). (Sources for mating system and size dimorphism: Cramp and Simmons 1981, and Dunning 1992)

not all species can count as independent observations. As an example, whole lineages may for historical reasons lack dimorphism in a given character. If, for example, a lekking lineage diverged from a single species that had arboreal displays and in which there was no advantage for larger males, all extant species, still displaying in trees, may lack sexual size dimorphism. The arboreal habitat, posed by the evolutionary history, constrains any selective advantage for large males and thus no dimorphism should be expected. A further example of the role of phylogenetic history is the sensory system used by a given lineage. The ancestral species could, for example, have been a nocturnal animal that had lost or never developed visual signals. Instead, the sensory system used for communication may have been vocal signaling. The extant lekking species, still nocturnal, could in this example be expected to display by means of auditory signals and there would be no reason to expect such a species to show elaborate color dimorphism. Such biases have been coined "sensory drive" (Endler 1992). As implied by these examples, lekking may not be a common selective regime. On the contrary, there are reasons to expect the opposite. As we have seen in chapter 3, successful males in different lekking species differ widely in the properties that make them successful. Given the great variability in lekking systems, it would be surprising if the same characters would be favored in all species.

Sexual selection is potent also in species with non-lekking mating systems. There are reasons to believe that most lekking lineages evolved from species with less extreme forms of polygyny (Höglund and Sillén-Tullberg, 1994). For example, in Calidrine waders, most species are monogamous or double clutching (the female lays a clutch with one male who incubates it, then leaves the male, mates with another male, and incubates the second clutch herself). All these species have a female-biased size dimorphism (fig. 6.1). However, four species in this subfamily have male-biased size dimorphism. These species are either polygynous or lekking. It is reasonable to assume that male-biased sexual dimorphism evolved in a polygynous, but non-lekking, common ancestor. Lekking as such has not caused the shift from female-biased size dimorphism to male-biased. The shift is probably brought about by the evolution of an ancestral polygynous mating system.

6.2 Different Comparative Approaches

SPECIES VERSUS OUTGROUP ANALYSIS

In an extensive review of size dimorphism in lekking birds, Payne (1984) concluded that in most species of lekking birds males tended to be larger than their conspecific females. However, Payne also found a number of exceptions. A study that used the outgroup comparison method (Ridley 1983) to control for the effects of phylogenetic relationships could not find that sexual size dimorphism and color dimorphism were more common in lekking clades as compared to non-lekking close relatives, although a simple comparison based on species as independent units showed the same results as those obtained by Payne (table 6.1; Höglund 1989). By contrasting the results on species and on clades it was suggested that common descent is important in explaining why lekking birds are dimorphic.

Could the negative result of the outgroup analyses be explained by reduced sample sizes, increasing the risk of a statistical type II error? An analysis of the change in the proportions of the presence/absence of dimorphism, when looking at species and clades respectively, suggests that for size, this is probably not the case. In table 5.1, 79% of all lekking species

Table 6.1

(a) Presence or Absence of Size and Plumage Dimorphism in Lekking and Non-Lekking Birds and (b) An Outgroup Comparison of the Presence or Absence of Size and Plumage Dimorphism in Lekking and Non-Lekking Birds

| | Sexual Size Dimorphism | | | | | Plumage Dimorphism | | | | |
| | Present | | Absent | | | Present | | Absent | | |
Mating System	N	%	N	%	P	N	%	N	%	P
(a)										
Lekking	69	79	18	21		55	62	33	38	
					.001					<.01
Non-lekking	13	50	13	50		8	31	18	69	
(b)										
Lekking	11	55	9	45		11	58	8	42	
					>.18					>.23
Non-lekking	6	40	9	60		6	46	7	54	

SOURCE: After Höglund 1989.
NOTE: P's are based on Fisher's exact test.

are size dimorphic whereas 55% of all clades are so. For plumage dimorphism the proportions are relatively unchanged. Thus, at least for size dimorphism, reduced sample size cannot be the whole answer, whereas for plumage it could have had an effect. But, even if the negative result were explained by the effects of reduced sample size, eight of nineteen lekking clades lack plumage dimorphism. This deficiency occurs often enough to make it more than an exception to a rule.

The result of the outgroup comparison shows that lekking seems to be correlated with size and plumage dimorphism in certain clades, and such clades often contain many species (fig. 6.2). Examples of such clades are lekking grouse with respect to size, some cotingas (especially the *Rupicola/Phoenicircus* line), lekking manakins with respect to plumage, and lekking birds of paradise. Other clades contain fewer species, and here lek-breeding species tend to be drab and the sexes equal in size. Examples of this kind are waders, parrots, tyrant flycatchers, and bulbuls. Yet other clades contain species where both lekkers and non-lekkers are dimorphic. Examples of this kind are grouse with respect to plumage, bustards, and weaverbirds. Two effects of phylogeny can thus be seen. First, lekking may have evolved in already dimorphic lineages, were dimorphism may have been promoted by sexual selection in non-lek systems, as exemplified by plumage in grouse, weaverbirds, and bustards. The dimorphism may in these groups have been taken to more extremes in lekking clades, for example in the grouse. Second, in some lekking clades size dimorphism has failed to evolve, as in manakins, tyrant flycatchers, and bulbuls, and in

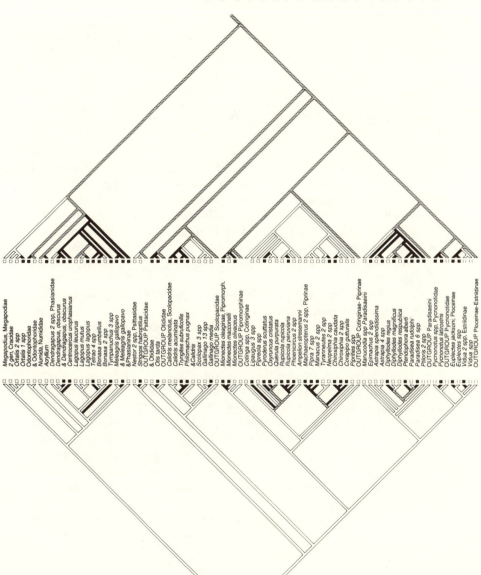

Figure 6.2 A compound phylogeny consisting of lekking taxa in the class Aves. An outgroup with a non-lekking state has been added to each of these monophyletic groups. (*Left*) The character mating system, consisting of two states, lekking (black) and non-lekking (white). (*Right*) Size dimorphism (as defined in Höglund 1989; black: males larger; white: no dimorphism or females larger; hatched: equivocal) has been reconstructed. Relationships between monophyletic groups follow Sibley and Ahlquist (1990). Lekking has most often evolved in already dimorphic lineages. (After Höglund and Sillén-Tullberg, in press)

Megapodius, Megapodiae
2 gen, Cracidae
Ortalis 2 spp
Ortalis 1 spp
Odontophoridae
& Odontophoridae
Numida, Numidiidae
Acryllium
Dendragapus 2 spp, Phasianidae
Dendragapus, obscurus
& Dendragapus, obscurus
Centrocercus urophasianus
Lagopus leucurus
Lagopus mutus
Lagopus lagopus
Tetrao 4 spp
Bonasa 2 spp
Bonasa umbellus
Tympanuchus 3 spp
Melagris gallopavo
& Melagris gallopavo
&Phasianinae
Nestor 2 spp, Psittasidae
Strigops habroptilus
OUTGROUP Psittacidae
& Otididae
Otis tarda
OUTGROUP Otididae
Calidris melanotus, Scolopacidae
Calidris acuminata
Tryngites subruficollis
Philomachus pugnax
& Calidris
Scolopax 3 spp
Gallinago 13 spp
Gallinago media
OUTGROUP Scolopacidae
Micronectes oleaginae, Pipromorph.
Micronectes macconelli
Micronectes olivaceus
OUTGROUP Pipromorphinae
Cotinga spp, Cotinginae
Lipaugus 3 spp
Pipreola spp
Pyroderus scutatus
Oxyruncus cristatus
Querula purpurata
Rupicola rupicola
Rupicola peruviana
Phoenicircus 2 spp
Ampelion stresemanni
Machaeropterus 2 spp, Pipirinae
Pipra 7 spp
Manacus 2 spp
Tyranneutes 2 spp
Neopelma 2 spp
Chiroxiphia caudata
Chiroxiphia 2 spp
Corapo guturalis
Pipirles spp
OUTGROUP Cotinginae- Pipirinae
Manucodia spp Paradisaeini
Epimachus 2 spp
Astrapia splendidissima
Astrapia 4 spp
Diphyllodes respublica
Diphyllodes magnificus
Ptendophora alberti
Paradisea rudolphi
Paradisea 6 spp
Parotia 3 spp
OUTGROUP Paradisaeini
Pycnonotus spp, Pycnonotidae
Pycnonotus latirostris
OUTGROUP Pyconotidae
Euplectes jacksoni, Ploceinae
Euplectes spp
Vidua 2 spp, Estridinae
OUTGROUP Ploceinae-Estridinae

others, such as snipe, both size and plumage dimorphism are lacking. We can think of at least three explanations for this, which we will discuss below.

The first explanation is that lekking is not associated with strong sexual selection in males and therefore does not promote dimorphism. Several lines of evidence suggest that this is not a valid explanation. As argued above (section 4.2), intense sexual selection on males is almost invariably found on leks (see also Payne 1984, Wiley 1991).

The second explanation is that monomorphic taxa have recently become lek breeding and there has not yet been time to evolve dimorphism. The fact that monomorphic clades often consist of a few species suggests that this may sometimes be the case. However, we feel that if there is any directional selection for larger size, there should have been ample time for almost all extant species to evolve dimorphism.

The last explanation is that in certain clades, sexual selection, even if potent, does not favor larger or brighter males but other characters instead, as we would expect if lekking is not a common selective regime. With regard to size, Höglund (1989) found that among clades that displayed on the ground and where male fighting is often observed, more have larger males compared to non-lekking clades. This makes sense, since in species that lek on the ground fights between males are common, and there are probably no premiums in being agile. On the contrary, in species that have arboreal or aerial displays, large male size may be selected against (cf. reasons for reversed sexual dimorphism in raptors, e.g., Cade 1960, Andersson and Norberg 1981). Furthermore, in table 3.1 we listed the studies that have revealed traits that give lekking males an advantage, and in only three species was large male size found to be associated with mating success.

With regard to plumage, dimorphism is expected only in species that use visual displays. In species that rely primarily on acoustic or olfactory communication we would not expect exaggeration in such cues. An environmental factor that could promote auditory signals is the light conditions in which a species is active. For example, in nocturnal species such as the great snipe, vocal signals may be more important than visual ones. Endler (1987, 1991, 1992) has shown how the color of male guppies changes with the light conditions for populations in different streams. In clear streams males tend to be brilliant whereas in poor light conditions males tend to be drab. That sexual selection favors alternative characters other than size and plumage can perhaps also explain why there are more species in clades in which plumage dimorphism is common. Slight plumage differences between local populations may promote reproductive isolation and speciation (Lande 1987).

The negative result of Höglund's (1989) study could be explained by the fact that only two aspects of sexual dimorphism were analyzed (size and plumage). Differences between clades in which factors that are promoted and prone to sexual selection are most likely explained by historic factors.

OUTGROUP ANALYSIS VERSUS INDEPENDENT CONTRASTS

Recently a comparative method—comparative analysis by independent contrasts (CAIC)—has become increasingly used in the analysis of comparative data that contain one or more continuous variables (Pagel and Harvey 1988). This method is based on the work by Felsenstein (1985). The logic behind it is that the value of a trait in a given set of species cannot be considered independent; instead, only differences (contrasts) that have emerged since the species diverged can be considered independent. Consider the four imaginary species (A to D) in figure 6.3. The difference $d1$ between A and B is considered independent since any difference between them must have arisen since the species diverged. Likewise, $d2$ as well as $d3$, which is at a higher node in the phylogeny, are independent contrasts (Harvey and Pagel 1991).

A recent analysis used a contrast method to reanalyze the patterns of sexual size dimorphism in lekking birds (Oakes 1992). The statistical test used paired tests comparing the means of the independent contrasts. This approach allows analysis of continuous characters and yet controls for phylogenetic effects. Oakes found that sexual dimorphism as measured by wing length was significantly larger in lekking birds (paired $t = 2.46$, $df = 16$, $P = .026$). However, this result could be biased by a relationship between sexual dimorphism and body size (see Clutton-Brock et al. 1977 for primates, Sæther and Andersen 1988 for grouse [Tetraonidae]). Analysis of the residuals from the relationship between male and female wing length also gave a significant result (paired $t = 2.62$, $df = 16$, $P = .019$; Oakes 1992), indicating that not all dimorphism can be explained by larger size. Oakes concludes that males of lekking birds, more often than non-lekking ones, are larger than their conspecific females.

How can the differences in the studies be explained? The outgroup comparison does not take into account the magnitude of dimorphism and therefore does not yield significant results. On the other hand, it is a powerful technique to be able to detect where in the phylogenies dimorphism originates. It also infers where monomorphism occurs and suggests in which clades to look for alternative targets of sexual selection. This could be illustrated by the fact that the outgroup method could detect a difference in sexual size dimorphism when only clades that display on the ground were analyzed (Höglund 1989). While, on the other hand, the analysis of the means of the contrasts allows analysis of continuous data and therefore is less prone to statistical type II errors, clades in which there is no dimor-

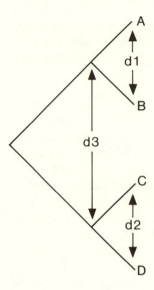

Figure 6.3 A schematic illustration of the calculation and use of independent contrasts. See text for explanation. (After Harvey and Pagel 1991, and Purvis 1991)

phism are swamped in the overall comparison. For example, the generic mean difference in male wing length in Pipridae (considered monomorphic in Höglund 1989) was 1.30 mm for lekking and 1.20 mm for non-lekking species. In contrast, the two comparisons in the family Paradisaeidae (considered dimorphic by Höglund 1989) were 29.95 mm versus 2.50 mm and 26.65 mm versus 14.20 mm, respectively. It is thus inescapable that in some groups there are large differences, and in some the differences between males and females are small, despite a lek mating system. In a test that takes magnitude into account, as does a paired t-test, a few very large comparisons may swamp more subtle differences. Differences among clades need to be explained. Recognizing the phylogenetic effects and the reason for why there are such effects is the only way one can do this.

Magnitude of Change or Number of Evolutionary Events?

We have reanalyzed Oakes's wing-length data (see also Höglund and Sillén-Tullberg 1994) with independent contrasts using the CAIC 1.2 computer package (Purvis 1991, Pagel 1992). In this analysis, we investigate whether positive contrast in wing-length differences between males and females (sexual size dimorphism) is correlated with mating system (lek or not). Thus we have one continuous variable (sexual size dimorphism) and one discrete character (mating system). If lekking is associated with greater sexual size dimorphism, we expect contrasts to be positive (the lekkers showing larger size dimorphism than the non-lekkers) rather than negative (the lekkers showing smaller size dimorphism than the non-lekkers). We used the same phylogeny as used by Oakes and the same categories for mating system with the following exceptions. In the first set of analyses we assumed the mating system of the wild turkey (*Meleagris gallopavo*) and the blue grouse (*Dendragapus obscurus*) to be non-lekking. Lekking in the wild turkey has been reported only once in one population (Watts 1968), and over most of its range it seems not to lek (Porter 1985). The mating system of blue grouse appears similar to an exploded lek in some studies (Lewis 1985), while others suggest a wider dispersion of the males (Bendell and Elliot 1967, Zwickel and Bendell 1972). In the second set of analyses, both species were included as lekkers. Three species of manakins considered to be lekking by Oakes (*Tyranneutes stolzmanni*, *T. virescens*, and *Neopelma chrysocephalum*) were considered to be non-lekking in all analyses. Oakes used Payne (1984) as the reference for the mating system of these three manakins. However, Payne used, among others, Sick (1967) and D. Snow (1963a). Sick refers to both *Tyranneutes* and *Neopelma* as "solitary" and Snow writes that "males display well apart from their fellows" (p. 554).

There were nineteen independent contrasts when we assumed that turkeys and blue grouse do not lek and twenty if they do. Ten and eleven contrasts were positive, respectively, and thus in these analyses there was no evidence of lekking birds having more sexual size dimorphism than non-lekking ones (sign tests, $P = .50$ and $P = .75$, respectively). Thus, whether or not the turkey and the blue grouse were included did not affect the results. Also, if contrasts were calculated assuming branch lengths proportional to the number of speciation events (Grafen 1989) or assuming all character change occurs at speciation events (punctuated equilibrium), the results were only marginally affected. We also controlled for the fact that dimorphism increases with size by analyzing dimorphism and female wing length with CAIC. We then obtained a corrected dimorphism (used in the analyses) as the residuals from the relationship of the contrasts between observed dimorphism and size (Höglund and Sillén-Tullberg, 1994). Six contrasts were positive in each analysis, and thus in these analyses there was also no evidence of lekking birds having more sexual size dimorphism than non-lekkers (sign tests, $P = .92$ and $P = .94$, respectively).

Thus we have two analyses (based on outgroups and independent contrasts analyzed by sign test): one that does not support the idea that lekking birds should be more dimorphic than non-lekkers, and one that supports it (based on independent contrasts analyzed by paired t-test). So, are lekking birds more dimorphic than non-lekkers or not? In one narrow sense the answer to this question is yes. However, we will argue that this depends heavily on what has happened several times in the evolution of lekking birds. Oakes used a test that takes the magnitude of a difference into account (a paired t-test). In some of his pairwise comparisons the difference between the lekking and non-lekking taxa is very large (two of the three independent comparisons within the grouse family, in the bustards, once in the waders, and in two out of three comparisons in the birds of paradise). The less impressive and even negative comparisons with less magnitude in other groups thus have less impact. In fact, only five out of seventeen pairwise comparisons show large differences, favoring the hypothesis that lekking birds should show more size dimorphism than non-lekking.

Neither the outgroup analyses nor the sign test of the independent contrasts takes magnitude into account, and in none of these analyses was there any indication that lekking birds usually showed more size dimorphism than non-lekkers. A test that takes magnitude into account is more powerful than one based on order. However, a comparative study should answer whether lekking leads to similar adaptations in independent evolutionary events. We argue that among birds, lekking has, in a few cases, indeed been associated with rather dramatic differences in size between the sexes. However, in many instances—in fact quite often in the evolutionary history of birds—lekking has not resulted in sexual size dimorphism. Thus it is fair to conclude that lekking is not always associated with sexual size dimorphism.

6.3 Social Selection

An alternative hypothesis for monomorphism despite intense sexual selection is social selection that affects both sexes (Trail 1990). Social selection is defined as direct competition through social interaction (Crook 1972, West-Eberhard 1979, 1983). According to this reasoning, both male-male and female-female competition during mating can favor the same characters in both sexes and lead to sexual monomorphism. Trail (1990) suggested a number of ways in which social selection may favor monomorphism in lek-breeding species. First, he suggested that low lek stability, caused either by male movement or high male mortality, may switch female choice of dominant males from cues correlated with dominance to direct assessment, and would therefore not favor exaggeration of male characters. Second, monomorphism may be favored because it may benefit low-ranking males to behave like females on the lek. By sneaking access to the lek, they can also get access to females. Females may also benefit by having a male appearance to avoid forced copulations and aggression (Geist 1974, Bleiweiss 1985, Robertson 1985). Finally, intense female-

female competition for particular males on the lek may favor elaboration of the same characters that favor males in male-male competition (see Otronen 1988 for a non-lekking example). Whether all these factors really lead to monomorphism remains to be analyzed in detail. The answers to such questions can be obtained only by detailed ESS modeling. It is, for example, not clear how monomorphism can be an ESS by having a selective advantage only in low-ranking males. Intuitively, such an advantage in low-ranking males would lead to the evolution of female-like appearance only in low-ranking males, and this may be the reason for delayed plumage maturation seen in some birds (Lyon and Montgomerie 1986, Grant 1990, Thompson 1991). However, such a process is unlikely to lead to complete monomorphism.

Social selection could be important through one or more of its proposed mechanisms in some species, in particular species such as the capuchinbird where both sexes have elaborate plumages (Trail 1990). However, some of the other examples suggested by Trail are not likely to be explained by social selection. In species where the sexes to a large extent are indistinguishable and drab, sexual selection on alternative characters apart from plumage and size is a more likely explanation.

Trail suggested that some populations of great snipe could fit the social-selection hypothesis. However, leks of this species are not unstable, as required by social selection, at least not in Scandinavia (Höglund and Lundberg 1987, Höglund 1989, Höglund et al. 1990a, Höglund and Robertson 1990a, Fiske 1994). This species is usually also highly territorial, even on rather small leks (Höglund and Robertson 1990a). Trail cited Lemnell (1978) as evidence of lack of territoriality which would promote direct assessment of male dominance. However, the studies of Höglund and co-workers were in the same area, and even Lemnell was able to draw territory maps on the two rather small leks he was studying. If lack of territoriality could be inferred from Lemnell's study, and this is doubtful, a small effect could be attributable to the small lek size. Female-female disruption has never been observed in great snipe, and there is no evidence of males behaving as females to gain access to the leks.

6.4 Summary

What is required from future comparative studies? It is necessary to explicitly state the specific hypothesis to be tested. This requires that dimorphism is defined and that more than a few aspects of dimorphism are studied. More detailed information on the precise selective agents, that is, female choice and male-male competition, is needed, and it is necessary to understand why in each species a character may confer an advantage. Comparative methods that allow analyses of continuous characters while still controlling for phylogenetic effects now exist (Brooks and McLennan 1991, Harvey and Pagel 1991). Recent developments in the methods of reconstructing evolutionary history of organisms have improved the phylogenies of birds and other animals (e.g., Hennig 1966, Eldredge and Cracraft 1980, Wiley 1981). Future studies of dimorphism in lekking birds and other animals are thus likely to unravel the relative importance of adaptation and constraint.

PART III   Lek Evolution

7

A Review of Hypotheses

7.1 Introduction

Though leks have evolved, they are not adaptations. Adaptations evolve because the genes for the trait have been favored by natural and/or sexual selection. Leks are the outcome of the behavior of many individuals in a population, and these behaviors are selected to maximize reproductive success at the level of the individual. Selection does not work directly on the mating system but on the behaviors that determine a mating system, for example on the genes that determine tendencies to join aggregations of other displaying males; when reviewing hypotheses on the evolution of leks, it is important to remember this. Leks evolve because the genetic programs of individuals are selected to make individuals behave in a way that makes lekking the observable outcome.

The general patterns of parental care and resource distribution reviewed in chapter 2 can somehow explain why leks occur in some species and not in others (see section 2.8). Ecologically, perhaps the most interesting aspect of leks, however, is the spatial clustering of males at specific sites. In this chapter a number of specific hypotheses aiming to explain spatial clustering are reviewed (see also Davies 1978, Bradbury 1981, Arak 1982, Bradbury and Gibson 1983, Vehrencamp and Bradbury 1984, Wiley 1991). A few of these, for example that leks have evolved to allow individuals to assess the population density for the good of the species, are clearly not compatible with modern views of natural selection. However, even after removing such hypotheses, a plethora of possible explanations still exists. Many of these probably have some validity for particular cases and all are not mutually exclusive.

Our reason for reviewing these hypotheses is mainly historical and for the sake of completeness. However, the different hypotheses differ widely in complexity and exactness: some are explicit mathematical models, others are verbal arguments, some address general mechanisms, and yet others suggest that specific selection pressures are responsible for the evolution of leks. In chapter 9 we will put the ideas we think are the most plausible into a general framework of formal models.

7.2 Predation Risk

One reason why some species mate on leks may be because the risk of predation on both males and females is lower when aggregating than during solitary display (Koivisto 1965, Lack 1968, Hjorth 1970, Wiley 1974, Wittenberger 1978, Oring 1982). While the costs and benefits of feeding in groups in relation to predation risk have been studied extensively both in theory and empirically (e.g., Pulliam 1973, Bertram 1978, Pulliam and Caraco 1984), there are no formal mathematical models on the risk of predation on leks. While this is no serious problem since such models are unlikely to differ from those derived for feeding groups, the empirical support for predation being lower on than off leks is more ambiguous.

The potential benefits of aggregating in relation to predation are several. First, if the number of attacks by predators does not increase in direct proportion to group size, the risk of predation for a particular individual decreases with increasing flock size (Hamilton 1971). Second, in flocks each individual can allocate more time to behaviors such as feeding, or on leks to display, and less time to scan for predators as flock size increases (Pulliam 1973). Thus, overall vigilance and the probability of detecting approaching predators increases with lek size (Kenward 1978, Bertram 1980). Finally, individuals in a flock may benefit by being in larger groups because large groups may be able to mob the predator and/or a large group may confuse an approaching predator (Wicklund and Andersson 1980, Treherne and Foster 1981, Götmark and Andersson 1984). Individuals in a large group will thus enjoy increased predator defense.

The risk of predation on leks indeed seems to be low. In one of the first attempts to quantify predation on leks, a thirty-five-year study of greater prairie chickens found only three successful attacks in 1,379 encounters between prairie chickens and the principal predator, the golden eagle (Berger et al. 1963; see also Oring 1982). Similarly, in sage grouse and in North American prairie grouse, the risk of predation on leks is generally very low (Hartzler 1974, Oring 1982, Gibson and Bachman 1992). The list can be extended: in our own studies of black grouse we have never observed a successful attack on lekking birds by the main predator, the goshawk, during five years of field work (Alatalo et al. 1991). Yet goshawk predation is the main reason why adult black grouse die in our populations. In great snipe, predation on the leks is also very rare (Höglund et al. 1992a). While a few great snipe males have been found dead on the leks with signs indicating they were killed by a raptor, no successful attacks by approaching predators have been observed. However, lekking males often ceased to display and hid or left the lek when predators approached (table 7.1). In

Table 7.1

Observations of Predators at Great Snipe Leks and the Subsequent
Behavior of Predators and Lekking Males

Species	No. of			Behavioral Response of Lekking Males		
	Obs.	Attack	Successful	Ignore	Hiding	Flying
Short-eared owl						
Asio flammeus	26	6	0	17	6	3
Hawk owl						
Surnia ulula	2	1	0	0	1	1
Hen harrier						
Circus cyaneus	3	1	0	0	1	2
Fox						
Vulpes vulpes	2	0	0	2	0	0
Total	33	8	0	19	8	6

SOURCE: From Höglund et al. 1992.

topi, leks are formed in places where the grass on the savanna is short and thus the risk of predation by lions may be low (Gosling and Petrie 1990). The same has been suggested for Uganda kob (Buechner 1961, Deutsch and Weeks 1992).

Even if low predation risk on leks is compatible with predation being a major selective factor in the evolution of leks, the critical test is if predation risk is lower on larger leks than on smaller ones, and, most conclusively, if predation on leks is lower than off leks. Few studies have been able to address these later questions. We will first turn to comparisons among leks of different size.

In túngara frogs, the per capita predation risk decreased with lek size (fig. 7.1; Ryan et al. 1981). Likewise, in cock-of-the-rock, males in larger leks suffered less predation than in smaller ones (Trail 1987). A recent study of a swarming chironomid midge also showed that the probability of predation decreased as swarm size increased (Neems et al. 1992; fig. 7.2). These studies thus suggest that an antipredator benefit of displaying in large aggregations could select for the habit of joining leks. However, in sage grouse, predator harassment and risk rates increased with lek size, contrary to the expectation of an antipredator benefit in larger leks (Bradbury et al. 1989a). The evidence from comparisons of lek sizes is thus contradictory.

The most detailed study on the risk of predation in a lekking species is from the Uganda kob (Balmford 1990). By comparing the number of male skulls of kob killed by lions on leks and in off-lek grassland, it was shown

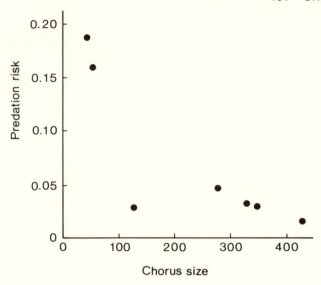

Figure 7.1 Risk of being predated by bats in relation to chorus size in the túngara frog. (From Ryan et al. 1981)

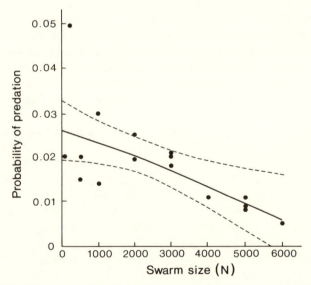

Figure 7.2 Probability of predation in relation to swarm size in a chironomid midge (*Chironomus plumosus*). The line shows the relationship: Probability of predation = .02 − .3 × 10^{-5}swarm size; dotted lines indicate 95% confidence intervals. Values are arcsine transformed. (After Neems et al. 1992)

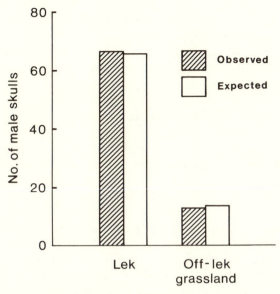

Figure 7.3 Observed and expected numbers of male Uganda kob skulls on leks and along transect lines in the surrounding savanna grassland. The density of skulls on leks and in grassland does not differ from what is expected by the density of living animals (χ^2 = 0.0031, df = 1, P > .95). (After Balmford and Turyaho 1992)

that in proportion to the numbers of kob present on and off leks, kob skulls were not found on the leks more often than in off-lek grassland (fig 7.3). Thus, in Uganda kob, lekking males were not safer from predators than non-lekking males. Lek animals were more vigilant and detected predators from longer distances, but these benefits seemed to be offset by lions concentrating their hunting efforts on the leks.

The most damaging evidence against predation being the only explanation for the evolution of leks comes from studies of lekking populations where there are no predators. In fallow deer, leks are formed in semicaptive populations kept in enclosed parks (Clutton-Brock et al. 1988, Langbein and Thirgood 1989, Appolonio et al. 1989a). While there is no predation in these parks at present, predation may be important in natural populations and may have been so in the ancestral stock of present park populations. However, in Petworth Park in England, an enclosed fallow deer population has been kept since the early nineteenth century, as evidenced by a picture of the fallow deer lek at Petworth by the English painter, J.M.W. Turner. Thus, in this population, lekking has persisted for nearly two hundred years without predators.

Even though predation affects the behavior and presence of individuals on leks (Tuttle et al. 1982, Trail 1987, Balmford 1990, Höglund et al. 1992a) and the geographical location (Gosling and Petrie 1990), the present evidence suggests that predation is not the sole explanation for why leks are formed in some taxa and not in others.

7.3 Information Sharing

In species that nest in colonies or roost communally, it has been suggested that a possible benefit of aggregation is that individuals can share information about where patchily distributed food is localized (Horn 1968). It has been hypothesized that leks may function in a similar manner (Emlen and Oring 1977).

In black grouse, gregarious foraging occurs and males may find food by following others when leaving the lek (de Vos 1979). However, this hypothesis cannot explain lekking in species that lack gregarious feeding. Instead, it seems more likely that gregarious feeding and possible benefits of information sharing are consequences of, and not a reason for, lekking in some species (Oring 1982).

7.4 Passive Attraction

In passive attraction explanations, females simply go to the lek providing the largest perceived stimulus (Otte 1974). In most cases this will be the site closest to a given female's home range. However, if a female is placed equidistant between two sites, she will prefer the one with more males since this site provides a larger stimulus than the smaller one. Males could thus benefit by joining other displaying males and therefore be able to attract more females. However, as shown by Bradbury (1981), a passive attraction model can never explain an increase in the number of visiting females on a per male basis. The number of females per male can, according to this explanation, at best be proportional to the number of males (an ideal free distribution) since the sum of two stimuli can never exceed the number 2. Even strict proportionality is unlikely since the value of every added stimulus is likely to fall with increasing group size. This is particularly the case for species that advertise their presence by the means of auditory stimuli. For simple reasons related to the physics of sound propagation, the value of an added ith stimulus is not directly proportional to the $i-1$ stimuli already present (Hassal and Zaveri 1979). On the contrary, sound stimuli, measured as sound power (dB), asymptotes quickly with increasing numbers of sound sources added.

Empirical tests specifically designed to test the passive attraction model are absent in lekking species. In dung flies, which have a mating system based on scramble competition for mates on the resource (the dung) where females oviposit, the number of males and females in each patch fit an ideal free distribution (Parker 1970). In this example, however, females are probably not attracted to the mating sites (patches of dung) by stimuli produced by the males, but by the state of the dung. Fresh dung is a more valuable food resource to the larvae than old dung. A further problem with the passive attraction model is that an ideal free distribution is also predicted by other models (see chapter 9). Also, in a wide range of lekking taxa, the per capita success of males increases with lek size (section 7.3), which is impossible in a passive attraction model. The theoretical and empirical evidence supporting the passive attraction hypothesis thus seems to be limited.

7.5 Habitat Limitation

An obvious reason for why animals are clustered is that the habitat in which a species lives has a patchy distribution. Several authors have stressed the importance of habitat limitations for the occurrence of leks (reviewed by Parker 1978b). It needs to be stressed that if habitat limitations explain all the male clustering and males space out in all the available habitat, we are not dealing with a lek (cf. chapter 1). However, it may be that habitat limitations can explain a part of the clustering and thus greatly influence where lek sites are formed.

Limited habitat may lead to male clustering in two different ways. First, males may require a distinct display habitat (e.g., a landmark) that may be limited in relation to the number of males in the population. Second, the breeding habitat of the females may occur in patches, and since the female distribution is clumped the male distribution will also be clustered.

In birds, there exists no evidence that display habitat is a limited resource. While some studies clearly suggest that males do prefer and select certain kinds of habitat when setting up a lek, most studies find that much of the selected habitat remains without leks (Koivisto 1965, Boag and Sumanik 1969, Snow 1974, Westcott, in prep.).

Breeding habitat, and hence the female distribution, is often clustered in space. In insects, male aggregations are commonly found near oviposition sites used by females (see references in Parker 1978b). For example, in some Diptera and Coleoptera, male mating aggregations can be found on cattle dung or carcasses (Pukowski 1933, Parker 1978a, Otronen 1990). Such sites, however, may not be similar to leks but are instead examples of resource-based leks. However, in many species it seems unlikely that males can control the patch, and therefore such systems bear similarities with

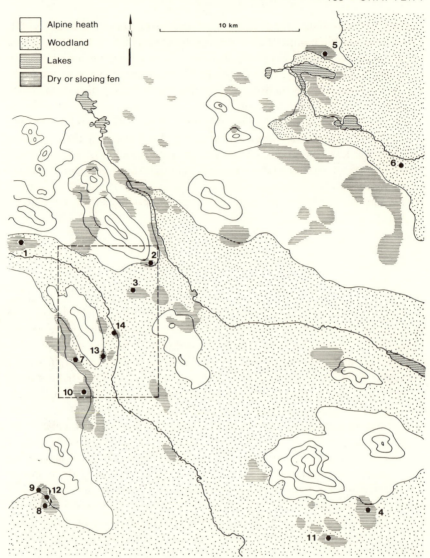

Figure 7.4 The location of great snipe leks in an area in western Sweden. The enclosed area was extensively searched for leks, and all existing leks were found. Outside, other leks may be present but not found. All leks but one are found on dry and sloping fens. (After Höglund and Robertson 1990b)

leks. Other encounter sites used by insects are hilltops and sunspots (section 1.5).

Females could also be limited in space by patchily occurring emergence sites, and males can be found swarming in such areas (Thornhill and Alcock 1983, Larsson 1989, Petersson 1989). Other factors that limit the availability of females are feeding sites and suitable nest sites. In birds, cock-of-the-rock leks are often found near steep cliffs in the rain forest, so-called *inselbergs*, in which females nest (Trail 1985a). Great snipe leks are found on wet and sloping fens in the Scandinavian mountains (fig. 7.4; section 1.4), and such sites may be patches in which females both feed and nest.

The patchy habitat distribution may thus determine why females are patchily distributed and would thus also explain why males cluster. However, while the importance of habitat limitation is widely recognized, some workers have challenged that male clustering on leks can be explained solely by this (e.g., Bradbury and Gibson 1983). It seems clear that the importance of habitat limitation may differ in different species. For example, in dung flies the habitat clearly limits the distribution of females and hence the males (Parker 1970). In other species, such as the great snipe, the habitat limitation may explain some of the clustering, but even within the preferred habitat males are found clustered further (Höglund and Robertson1990a). Thus the degree of male clustering could be thought of as a continuum with respect to habitat limitation. In some species habitat limitation is the sole or most important factor governing female and hence male distributions, while in others habitat limitations may be of less importance.

7.6 The Hotspot Model

One of the most discussed lek models during the last decade has been the hotspot model (Bradbury and Gibson 1983, Bradbury et al. 1986). In this model males settle in those sites where the probability of encountering many females is high (see also Lill 1976, Emlen and Oring 1977, Payne and Payne 1977). Hotspots could be determined by nondefendable resources to which females are attracted such as food, nest sites, and water holes. But even in the absence of resource gradients, hotspots could arise from overlapping female home ranges (see below; Bradbury et al. 1986).

The clustering of males is thus determined by the clustering of females, which in its turn could be determined by the clustering of resources. Is lekking always explained by the underlying resource distribution? Bradbury and co-workers (1983, 1986) tried to answer this question by modeling hotspots without considering resources. They argued that in species in which females have large and overlapping home ranges there will be some areas in which males encounter more females than in others.

Model simulations of this process assume circular female home ranges with a fixed radius in each run (Bradbury et al. 1986). Each run starts with a random dispersion of female home-range centers onto a surface (the habitat). The surfaces were either unbounded or had a fixed boundary, but both approaches gave similar results. The points of overlap on the surface are by definition hotspots. Males were then allowed to settle sequentially on the surface by examining fifty randomly selected locations. For each location, the expected number of matings was calculated at the time of settlement, and the male settled at the location with the highest expectation. In the basic model, each female encompassing this site was worth $1/N$ copulations to a potential settler, where N is the number of males at the site. This means that for the second male who settles, if the location of the first male contains more prospective copulations than alternative sites, he joins the first male despite the devaluation. If, on the other hand, a female home range outside the first site (unaffected by devaluation) has more prospective matings, the male settles in this site. The general outcome of these simulations is that males will tend to cluster on the points of greatest female home-range overlap, or greatest female density, and that female home-range size accounts for 25%–30% of the variance in male clustering (fig. 7.5). The mathematics and theoretical framework of this and other game theory models of leks are presented in chapter 9.

That leks are sited on the points of greatest female density is indirectly supported in some species of birds. For example, in the ruff, leks are often situated near small ponds that females probably visit for feeding and drinking (Höglund et al. 1993). In the Costa Rican rain forest, four species (a tyrant flycatcher, two species of manakins, and a hummingbird) set up leks at roughly similar places from year to year (Westcott, in prep.). This suggests that there are two environmental explanations for the location of leks. First, the topography of the landscape may funnel females into certain areas (leks are often in valleys separated by ridges). Second, some of the species have overlapping diets and thus they have similar resource needs. If resources occur at certain places, the males will set up their leks there. In any case, the results render at least limited support to the idea that males set up leks where the chances of encountering females are high.

If studies of birds seem to fit with the hotspot idea, studies of lekking ungulates do not. Studies of three species of lekking ungulates (reviewed by Balmford et al. 1993 and Clutton-Brock et al. 1993) indicate that the degree of clustering generally increases rather than decreases with increasing population density. Furthermore, census data on Uganda kob suggest that this species is characterized by lack of sites where female density is predictably high (Deutsch 1994). It is unclear in the ungulate examples, however, if the censused females are sexually receptive. If this is not the case, the census data are irrelevant in testing the hotspot model. The occurrence of leks in high-density populations, however, goes against the hotspot idea as formulated in the model by Bradbury and co-workers (1986).

Other results of hotspot modeling amenable to field tests have been suggested in field tests. First, female home-range size affects male clumping.

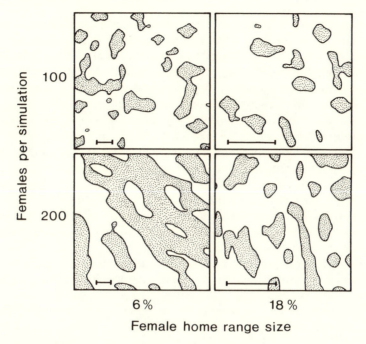

Females per simulation

100

200

6 %   18 %

Female home range size

Figure 7.5 Simulated contour plots of two hundred males settled on simulated female dispersions. Contours enclose sites with the top 75% male density. Bars indicate female home-range diameter. Percentage of area shaded: 18.7% (*top left*), 11.5% (*top right*), 50.9% (*bottom left*) and 27.2% (*bottom right*). (After Bradbury et al. 1986)

This means that with larger female home ranges, there are more males per site and thus longer distances between sites. Second, the male settlement rules produce sites that may be less than one female home range apart. This leads to the third finding—that females should visit several leks if the hotspot model applies. Finally, the clustering of male territories will decrease with increasing population density.

Direct empirical tests of the hotspot model should ideally include data on the size and overlap of female prenesting home ranges and detailed data on how males settle according to female distribution. However, since the basic model by necessity has to simplify matters, the basic assumptions in the model are rarely met in nature. In this model there are no additional factors such as habitat gradients, differences in male competitive ability, differences in female experience, and motivational status that may confuse the picture. For example, in migratory birds there is considerable difficulty in defining what a prenesting home range is (Höglund and Robertson 1990a). Field workers have therefore turned to indirect tests of the predictions of the model. These tests are indirect because the predictions are not exclusive to the hotspot model.

One indirect test of the hotspot model is that leks may be found less than one average female home range apart. Three studies have found that an average female home range encompasses more than one lek, in accordance with the hotspot model. In sage grouse (Bradbury et al. 1989b) and Lawe's parotia (Pruett-Jones 1985 in Bradbury et al. 1986), prenesting homeranges were larger than mean inter-lek distances. Furthermore, in greater prairie chickens, an indirect measure of female home-range size (nest to lek distances) showed that 74% ($N = 89$) of the females nested closer to a lek different from the one on which they were observed mating (Schroeder and White 1993). In two other studies on capercaillie (Wegge and Rolstad 1986) and great snipe (Höglund and Robertson 1990a), leks were further apart than predicted by female home ranges. Evidence from comparisons of female home ranges and inter-lek distances is thus contradictory.

Another prediction from the hotspot model is that females should visit more than one lek before breeding. All studies cited above found such a pattern in accordance with the hotspot model.

It has been suggested that the two predictions above could be used in discriminating between the hotspot model and models that assume female choice for lekking males (Bradbury 1981, Bradbury and Gibson 1983). However, this is based on assumptions of female-choice models that appear limited (we will discuss these assumptions in detail when discussing female-choice models below). Tests based on data of inter-lek spacing, female visiting patterns, and female home-range size thus at best render limited support to the hotspot model (Gibson et al. 1990).

Another indirect test of the hotspot model is to compare female home-range size in a range of lekking and non-lekking species. In grouse, antelopes, birds of paradise, and waders there is a negative correlation between male territory size and female home ranges (chapter 2; fig. 2.4), which provides evidence that the home ranges of lekking species are larger than the home ranges of non-lekking species (Bradbury, Gibson, and Vehrencamp in Krebs and Harvey 1988). Similarly, in six sympatric manakins (Pipridae), the home ranges of three lekking species were larger than in two explosive lekking species that had home ranges larger than those found in a species having single territories. Furthermore, home ranges of lekking species were found to overlap by 65%–75% (Théry 1992).

The causality of this last prediction can be questioned, however. Are female home ranges larger in lekking species because large ranges lead to leks or because the evolution of leks for other reasons means that females have larger ranges? Bradbury and co-workers suggested that tests of this prediction should use prebreeding home ranges (Bradbury and Gibson 1983, Bradbury et al. 1986). It is unclear, however, whether prebreeding home ranges can be considered independent of female ranges during breeding.

A final indirect test of the hotspot model is to study the number of mating females on leks in relation to the number of males present. If no other process than males settling with respect to female density determines the number of males in a site, we would expect males to distribute themselves in proportion to the number of females present. The distribution of males would thus follow an ideal free distribution (Bradbury et al. 1990b). As we will return to below, an increasing body of evidence suggests that males on leks compared to those displaying solitarily, and males in larger leks compared to smaller ones, receive a disproportionate number of the matings observed. The distribution of males thus does not appear to be ideal-free. However, this does not disprove the idea that males should settle in sites of high female traffic, only that other processes in addition to the hotspot mechanism are operating.

A final criticism of the hotspot model is that the simulations by Bradbury and co-workers do not generate very impressive clustering. For example, in the simulation that yielded most male clustering, 75% of the males were found in 11.5% of the habitat (Bradbury et al. 1986). In contrast, real leks take up less space: dividing the area covered by leks by the area of suitable habitat in three species of lekking birds gives estimates of leks covering less than 1% of the available area (black grouse, 0.12%; great snipe, 0.03%; ruff, 0.04%). Even if the estimates for real species are not directly comparable to model simulations, real species seem to be more clustered than predicted from the simulations. It is possible that in nature, females are not randomly dispersed but for ecological reasons are confined to certain areas. In any case, the simulations of the hotspot model do show that male clustering could be generated in the absence of ecological factors; they also give insights to how female range patterns can affect male spacing. In real life it is likely that hotspots are generated both by ecological resources determining female densities and by the degree of overlap in female home-range use.

7.7 Increased per Capita Mating Success on Larger Leks

There are two distinct groups of models that rely on a female preference for the evolution of leks. First, in what may be called preference for clustered males, females for some reason prefer to mate in leks. We will review suggested reasons for such a preference in detail below.

Whatever the reason for the female preference, there is a clear prediction from these models: for leks to evolve, the per capita success of males on leks must be higher than during solitary display. It is therefore of interest to investigate whether the per capita male mating success increases with lek size, even though this can also arise for reasons other than female preference (chapter 9).

Second, in the hotshot model, females have preferences for certain males but not for clustered males per se. Leks arise as an epiphenomenon because less-preferred males join the preferred to parasitize on their attractiveness. The hotshot model makes no clear prediction as to whether the per capita success of males should increase with lek size, but this ought to follow if the most attractive hotshot is surrounded by relatively more males.

EVIDENCE FOR INCREASED PER CAPITA SUCCESS ON LARGER LEKS

What is the evidence of an increased per capita success of males with increasing lek size? Four recent studies on insects show an increased per capita male success in larger aggregations. In *Drosophila conformis* larger leks had higher female visiting rates (fig. 7.6; Shelly 1990), and in the stink bug *Megacopta punctatissimum* females courted males in aggregations more often than solitary males (Hibino 1986). In the behaviorally sex-role-reversed dance fly *Empis borealis*, where females lek, average female mating success increased with the size of the swarms, and visiting males mated more often in larger swarms (Svensson and Petersson 1992). Finally, male mating success increased with male aggregation size in a Coreid bug (Nishida, in press). A somewhat more complicated pattern was found in the midge *Chironomus plumosus*. Male mating success was highest in the smallest swarms and decreased with increasing swarm size until swarm size was about four thousand animals. After this size, male success increased; thus larger swarms, also in this species, attracted more females (Neems et al. 1992).

In birds and mammals three studies have clearly shown an increased per capita success, whereas three other studies show ambiguous results. In the Uganda kob, mean ejaculation rate per male increased with lek size (Balmford 1990, but see Deutsch 1991). These results agree with our own study of black grouse in which larger leks had more female visits, more copulations per male, and in which females were more likely to mate in larger leks (fig. 7.7) (Alatalo et al. 1991). The latter result indicates that females really prefer to mate in larger leks.

The ruff is a lekking wader where males occur in two separate behavioral morphs: independent males that, depending on age and condition, defend territories on the leks; and satellites that do not defend territories but parasitize the reproductive effort of independent males (Hogan-Warburg 1966, van Rhijn 1991, Höglund et al. 1993). In this species the per capita success of independents increased with lek size (fig. 7.8). It should be stressed that in all these examples, the largest lek sizes were about 10–30 males. In the black grouse example, average male mating success seemed to level off after about a lek size of ten males, which was also the case in the kob.

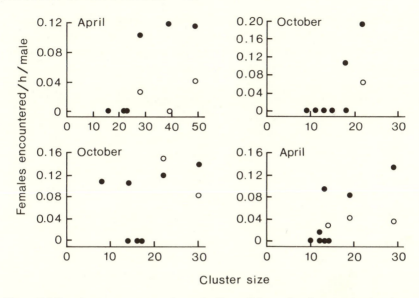

Figure 7.6 Female encounter rates for male *Drosophila conformis* occupying low leaves on the bushes where the lek was situated (filled dots), and high leaves (open dots) in differently sized clusters. (From Shelly 1990)

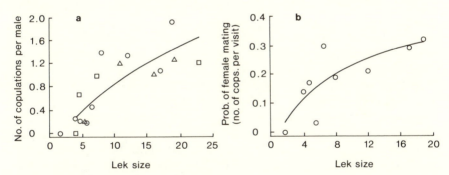

Figure 7.7 (a) Number of copulations per male (r_s = .82, N = 17, P < .01), and (b) the probability that visiting females mate in relation to lek size in black grouse (r_s = .0.85, N = 9, P = .01). (After Alatalo et al. 1992)

In the sage grouse, where leks can be very large (up to about two hundred males), female visiting rate did not increase on a per male basis with regard to the whole range of leks studied (Bradbury et al. 1989b). However, the visiting rate increased sharply, up to about 25–50 males, and then leveled off. Also in Uganda kob, there was no per capita mating success increase on very large leks (Deutsch 1992). In prairie chickens, the data suggest that male success does not increase with lek size (Hamerstrom and

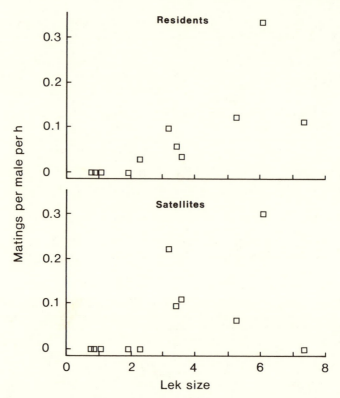

Figure 7.8 (*Top*) Number of matings per resident male $r_s = .91$, $N = 11$, $P < .001$), and (*bottom*) number of copulations per satellite male in relation to lek size in ruff ($r_s = .56$, $N = 11$, $.05 < P < .10$). (After Höglund et al. 1993)

Hamerstrom 1955). However, in this case it is doubtful whether large and small leks were sampled at the same time in the breeding cycle, and the data for large and small leks may not be comparable. For example, large leks (> 26 males) were sampled on fourteen mornings and a total of only four copulations were seen. The most likely explanation for this is that leks of this size were not sampled during peak mating.

An increased per capita male success can be explained without female choice (we will show why this is so in chapter 9). However, if females indeed prefer to mate in larger leks, males in larger leks will obviously benefit. There are two field experiments on birds and one on an insect which suggest a female preference for large leks, while yet another insect study finds no preference at all. First, offering black grouse females a choice between groups of six or three male models resulted in sixteen visits to the six-male group and three visits to the three-male group (Kruijt et al. 1972). Lank and Smith (1992) analyzed these data statistically and found

Table 7.2

Proportion of Female Ruff Visits to the Aviary with More Males When Given
the Choice of Two Male Groups.

	Females Only	Mixed Flocks	All Female Visits	Males Only	All Male Visits
TRIALS					
Observed	.84 (19)	.86 (7)	.87 (23)	.74 (22)	.76 (21)
Expected	.69	.66	.66	.66	.69
P more visits to larger group	.002	.063	.001	.017	.013
P per capita male benefit	.122	.242	.022	.283	.315
VISITS					
Observed	.78 (37)	.82 (11)	.79 (48)	.65 (54)	.68 (65)
Expected	.67	.65	.66	.66	.66
P more visits to larger group	.001	.033	.001	.020	.002
P per capita male benefit	.090	.202	.039	.490	.416

SOURCE: Modified after Lank and Smith 1992.

NOTES: In *trials* all visits during a bout of observations are lumped and in *visits* all female visits are regarded as independent observations. *N*-values are in parentheses.

a preference for the larger group (binomial test, $P = .002$), but the data were not sufficient to show a higher visiting rate on a per male basis (null hypothesis that $P = .67$, $q = .33$, binomial test, $P = .095$, one-tailed). Second, in an experiment with ruffs, two aviaries that varied in the number of males inside them were put 35 m apart in the field (a distance which, in the case of ruffs, is a short but not unreasonable distance between real leks) (Lank and Smith 1992). Pooled data from two years showed that both visiting males and females showed a preference for the larger group (table 7.2), but only females showed such a strong preference that males in the larger group were favored on a per male basis (table 7.2). Third, in an experiment with caged male bladder cicadas, more females per male were attracted to cages with more males than found naturally (Doolan and MacNally 1981). Finally, however, although male crickets were found to be aggregated in the field, no phonotactic preferences by females for aggregated over isolated loudspeakers that broadcasted male song were found (Cade 1981).

In conclusion, an increasing body of evidence from a wide range of animals seems to suggest a male advantage of displaying in groups as compared to solitary display or in very small groups. However, in studies of very large aggregations, no per capita advantage for the attending males seems to be commonplace. This is to be expected, since the attractiveness to females of ever-increasing aggregations is likely to level off, and thus the advantages to attending males will follow a relationship of diminishing returns. Furthermore, some studies do not find a per capita male advantage when comparing leks of different sizes.

REASONS FOR FEMALE PREFERENCE
OF AGGREGATED MALES

Assuming that increased per capita success comes about through female preference for mating in larger leks, what are the possible reasons for females showing such a preference (table 7.3)?

The first reason relates to predation. Females may prefer clustered males because mating with such males entails low costs since clustered males will deflect predators away from the nest sites (Wrangham 1980). Nesting success is thus predicted to increase with distance from the lek site. Phillips (1990) extended this idea by not only invoking males as decoys but also as sentinels for females. This hypothesis thus predicts that nesting success will be highest at some intermediate distance from the lek site. Even if Phillips published some data that could be interpreted as evidence of such a pattern, this is an unlikely explanation. Rarely does the same predator strike both adults and nests. For example, in black grouse, goshawks are the principal predators on adults (Alatalo et al. 1991), whereas nests are depredated by Corvids and foxes (Willebrandt 1988). Thus predator deflection could not work in this case. Females could prefer males on leks because such sites are safer from predation (Wiley 1973, Wittenberger 1978). However, as argued above, the evidence of leks as safe sites is somewhat ambiguous (section 7.2).

Males on leks could be preferred because clustered males allow females an easier comparison of males and reduced search costs than if males are spaced (Alexander 1975, Bradbury 1981). Also, the fact that males are aggregated may allow females to compare males in competitive interactions directly. While they are compelling, these issues have not been directly tested.

Another reason for a female preference of males in larger leks is that female preference for larger leks can evolve and be maintained without any direct selective advantage (Queller 1987). This is similar to how any sexually selected male trait could evolve under female preference (section 3.5).

The final reason for female preference of clustered males could be that males on large leks provide females a better breeding opportunity. First, the absolute quality of the best male could be better in the largest site (Beehler and Foster 1988). Second, the average male quality could be correlated with lek size (Alatalo et al. 1992, Höglund et al. 1993, Sutherland et al., subm.). Note that male quality in both of these explanations need not nec-

Table 7.3

Hypotheses for Female Preference of Large Leks

Hypothesis	References
Least costly males, predator deflection	Wrangham 1980, Phillips 1990
Reduced predation	Wiley 1973, Wittenberger 1978
Easier mate comparison	Alexander 1975, Bradbury 1981
Choice of better males—hotshots	Arak 1983a, Beehler & Foster 1988, Höglund & Robertson 1990b
Choice of better males—average quality better	Alatalo et al. 1992
Indirect selection for lek size	Queller 1987

essarily mean genetic quality even if this is possible. The evidence for these two possibilities relates to the discussion of the hotshot hypothesis and the ideal free distribution of unequal competitors (sections 7.8 and 9.2). At this stage it is sufficient to say that both explanations are possible.

7.8 The Black Hole Model

The most recent suggestion for the occurrence of leks is the black hole model (Stillman et al. 1993, Clutton-Brock et al. 1992). In this model there is no female preference for particular kinds of males or leks. Rather, females are indifferent about whom they mate with as long as the mating and courtship do not cause them any harm. If mating and related activities are dangerous to females and some kinds of males are more dangerous than others, females will tend to move before mating, which is one basic assumption of the black hole model.

For example, mating with young and inexperienced males may cause harm because an inexperienced male may not yet have acquired the skills of mating and thus cause the female injury. Another reason for avoiding inexperienced males, and perhaps a more plausible one, is that inexperienced males may not be able to fend off competitors. If disrupted and harassed matings are dangerous, it may pay a female to avoid such males.

If sexual harassment is a serious cost, it follows that sexually receptive females should often avoid certain males. Females thus not only enter male territories but may also leave them without mating. Eventually the female has to mate and will do so with the male in whose territory she happens to be present.

Given the above scenario, the optimal male nearest-neighbor distances can be calculated. In simulations it can be shown that if females behave as described above, males should cluster together. The reason for this is that even if females also leave males when males are clustered, the probability that males can retain females is much higher if the males are clustered instead of solitary. Thus, like stars in space, females are sucked into black holes.

In the next chapter we describe this model in greater detail. Some support for the model has been claimed for fallow deer leks (Clutton-Brock et al. 1992) and a marine iguana (F. Trillmich, pers. comm.).

7.9 The Hotshot Model

The hotshot mechanism suggests that females prefer to mate with attractive "hotshot" males and thus search the habitat for them. Unattractive males simply join this hotshot male to parasitize on his attractiveness. Thus, the clustering of males would be a result of the hotshot being attractive not only to females but also to less attractive males (Arak 1982, 1983a, 1988b, Beehler and Foster 1988, Höglund and Robertson 1990b).

This explanation was first proposed by Arak (1982), but the term "hotshot hypothesis," under which this mechanism is best known, was suggested by Beehler and Foster (1988). Other models of similar mechanisms have been developed for host-parasite interactions and food theft. These models are often referred to as kleptoparasitic or producer-scrounger models and do not explicitly treat lek systems (Parker 1984a,b, Barnard and Sibly 1981, Sibly 1984, Parker and Sutherland 1986). Beehler and Foster (1988) proposed the hotshot explanation as an alternative hypothesis to other proposed explanations for lek evolution, in particular the "female preference model" as stated by Bradbury (1981) and the "hotspot" model of Bradbury, Gibson, and Tsai (Bradbury and Gibson 1983, Bradbury et al. 1986).

Beehler and Foster (1988) criticized earlier models and in particular what they felt as a hidden assumption in them, namely that the most important selective pressure acting on males was mediated by female preference. Instead they advocated the importance of male-male competition, which they argued would inevitably lead to the evolution of stable dominance hierarchies on leks. However, we feel some of this criticism is unwarranted. First, while the Bradbury (1981) version of the female choice model indeed relies on female preference, the hotspot model does not assume female choice. It assumes that females mate within their home range, but that male success could theoretically be entirely shaped by male-male competition. Second, we fail to see the logic behind the male-male competition/ dominance argument in the hotshot explanation. Furthermore, the hotshot idea seems based on a female preference argument and could, in our opinion, work entirely without male combat.

To follow these arguments, what did Beehler and Foster actually say? They begin with a dispersed population where males defend courts but not resources. Males cluster as an effect of some males being preferred by females and some not. Unpreferred males should thus gain by becoming parasites on preferred males. "Ultimately, males cluster into leks because of the 'magnetic' effect of mating skew" (p. 208, Beehler and Foster 1988). Mating skew comes about because (1) "intrasexual physical and psychological aggression [prevents] some males [from] holding display courts"; (2) "some court holders are chosen by several females"; (3) "unsuccessful males abandon their courts"; (4) "conservative mating tactics [are adopted] by females"; and (5) "hotshot[s] begin attracting attendant males."

Let us examine each of these five steps. Arguments 2, 3, and 5 are similar to other versions of the hotshot model (Arak 1982, 1983a, 1988b, Höglund and Robertson 1990a). However, while it may be that some males are prevented from holding court because of physical and psychological aggression (1), remember that a first assumption was a dispersed population that is not defending resources. Disrupting males thus presumably have to spend time traveling from their own courts to the sites of their competitors. This is time that has to be traded against advertising for females at their own courts. Furthermore, the argument that females should adopt conservative mating tactics (4), which basically means anything but selection of phenotype, does not follow logically. Females may have adopted conservative mating patterns in some species, but we fail to see why this must follow from male clustering around hotshots. Turning the argument around, nonconservative female mating patterns could also lead to hotshot clustering.

General predictions from the hotshot model (excluding points 1 and 4 above) are as follows: evidence of kleptoparasitism should be found on leks; there should also be clear differences in mating success explained by differences in attractiveness and/or dominance status; for reasons given in chapter 9 (section 9.2) leks should be unstable. Especially low-quality males should be expected to change lek sites. Beehler and Foster (1988) listed some other predictions that relate to removal experiments of males. Removal of the best male should result in decreased visitation rates and loss in attendance by subordinates. Removal of several hotshots should result in increasing disruptions and fighting rates and eventually the breakdown of leks (assuming stable dominance hierarchies are prevalent). Removal of subordinates would have no effect on female visitation rates.

Few field studies have addressed the predictions from the hotshot explanation. In natterjack toads, however, there is ample evidence of kleptoparasitism. Large males predominantly adopt an advertising mating strategy by announcing their readiness to mate by calling. Females are passively attracted to male calls and when given choices move to the call that is loudest. Since sound pressure levels (dB) of male calls are positively correlated with male size, large males have a mating advantage over small males.

Small males are thus often found adopting an alternative sneak strategy by adopting a silent low posture next to calling males and trying to intercept females on the way to the caller (Arak 1983b).

In great snipe, males attract females by repeated display behavior, and the rate of display by any given male is positively correlated with male success (Höglund and Lundberg 1987, Höglund and Robertson 1990a). Attractive males end up in the center of the lek presumably because less attractive males compete for positions close to attractive ones. When peripheral males are removed, such territories are instantly taken over by neighboring or floating males, whereas removal of central birds results in reallocation of the lek. Similarly, peripheral birds attack dummies accompanied by playback calls broadcasted within their territories, whereas central birds are indifferent to dummy males (Höglund and Robertson 1990a). These results could thus be interpreted as evidence of kleptoparasitism. Also, in this species, dominance is positively correlated with male success (Höglund and Robertson 1990a). However, this result is probably an effect of attractive birds being visited by more females and hence attacked more often by less attractive males. Since dominance is clearly site asymmetric, all birds are dominant within their territories (Höglund et al. 1990a; section 3.2); therefore attractive birds are probably scored as dominant because they are more often attacked within their own territories. The last conformity with the hotshot idea in great snipe is the movement of unsuccessful birds. In this species, successful, and possibly highly competitive birds are site faithful, whereas unsuccessful birds do not stay on the leks as long and do not return at the same rate between years as do the successful birds (Höglund and Robertson 1990b).

Robel and Ballard (1974) removed males from prairie chicken leks. The subsequent changes of male territories could be interpreted as male reshuffling around other dominants. Furthermore, overall mating success of the lek decreased as the most dominant birds were removed, indicating that females moved to other leks. However, this experiment lacked controls and was not properly replicated.

While there is some evidence of kleptoparasitism on leks, we feel that the hotshot mechanism cannot be the sole explanation for lekking. It is evident that if lekking is simply a consequence of kleptoparasitism and females are attracted to a site only by the presence of one or several hotshots, it would pay them to move if they take the females with them. To explain why hotshots stay, either site limitations or tradition has to be invoked. However, if either of these were the case, the hotshot argument breaks down to an argument that is similar to other hypotheses of leks. Males are found where they are for the same reasons that bring many females to any given site.

7.10 Summary

In this chapter we have reviewed the proposed hypotheses as to why leks occur. The ideas are summarized in table 7.4. While a few of these can be considered as unlikely explanations (information sharing and passive attraction), the others are supported by evidence, at least to some extent. All of these remaining explanations may be important. However, their relative importance is likely to differ in different kinds of animals, and no single idea is likely to provide a sole explanation for lekking.

The reader may be disappointed by such a weak conclusion at this point. In the next chapters we propose a way to think about lek evolution and mating system evolution in general, which may be more revealing than trying to fit all species into a single hypothesis. Instead of trying to test the various hypotheses about leks with field data from different species, we feel it is important to try to understand the selective forces that act on males and females in different species. Individuals in all species behave in ways

Table 7.4

Summary of Reasons for Lek Formation

Theory	Mechanism	Conclusion
Display habitat limited	More males than display sites	Important in some systems; further clumping needed if leks
Clustered males avoid predation	Selfish herd, dilution, deterrence	Not likely as a single explanation
Clustered males forage more efficiently	Information sharing	Unlikely explanation
Leks have wider detection range	Passive attraction	Unlikely on theoretical grounds
Hotspots	Overlapping female home ranges or clustered resources	Cannot explain all clumping
Females prefer clustered males	Active attraction	Probably important in some systems
Black holes	Sexual harassment causes female movement and males retain more females when clustered	Most likely in relatively immobile species
Hotshots	Secondary clustering around attractive males	Unlikely to explain initial clustering

that maximize their reproductive success. Given the differences in the ecology, life histories, and habitat of different species, it should not be surprising if the occurrences of leks in different species can be explained by different factors. It is therefore probably more rewarding to try to understand the selective forces that act on spacing, sexual advertisement, and other behaviors that are likely to affect the reproductive success of animals in a given population. The resulting mating system can be seen as the net result of the behavior of all animals in the population, and sometimes the net result is leks.

8   Intraspecific Variation

8.1 Introduction

There is a long tradition in animal ecology of using interspecific comparisons to understand the variation in and evolution of animal mating patterns (e.g., Crook 1964, Emlen and Oring 1977, Clutton-Brock 1989). The basic approach has been to look for similarities in ecology that may explain similarities in mating systems across different species. While using this approach, animal mating systems have often been regarded as relatively static and species specific. However, recent studies of many animals have revealed a plasticity in the mating systems within the same species, suggesting that different selection pressures may shape different optimal mating behaviors in different populations (Lott 1991).

The mating system of a population can be seen as the summary of the behavior of the individuals in a given population. The costs and benefits both to males and females need to be analyzed because what is best for males may be in conflict with what is best for females (Oring 1982, Vehrencamp and Bradbury 1984). Furthermore, different kinds of individuals of the same sex may have different optimal mating strategies. There are well-documented cases that factors such as size and age influence the mating strategies adopted by individuals (Gross 1991).

Thus leks are unlikely to be explained by a species-level approach. Instead, a more useful approach is to examine the selective pressures that cause males to aggregate on the level of the individual.

In this book one of the main questions has been: Why do leks exist? A simple answer is that leks exist because males in populations of some species achieve higher fitness in aggregated than in solitary display. This answer is unsatisfactory, however. A deeper understanding of leks includes knowledge of the mechanisms and circumstances that generate a selective advantage of lekking. These mechanisms and circumstances are likely to vary not only among species but perhaps also among populations of the same species. Leks, as defined in this book, are aggregations of males visited by females primarily to mate. In the previous and following chapters, we review a range of models all of which may explain male aggregation. At this stage it would be premature to rule out any except perhaps a few of the models, as only a few specific tests are yet available. Furthermore, if some models can be ruled out in a given species, it does not mean that they cannot be applied to studies of other species.

Species that show a range of mating patterns within the species are likely to be important examples of the factors that cause males to aggregate. Comparisons of populations of "plastic" species can hopefully contribute to identify the ecological circumstances that cause males to lek or to use other spatial distributions.

8.2 Intraspecific Variability

THE CASE OF THE FALLOW DEER

One example of a plastic species is the fallow deer. It was formerly described as a species in which males were territorial with rather dispersed territories (Espmark and Brunner 1974, Chapman and Chapman 1975). Some recent papers have reported that males exhibit nonterritorial harem defense (Schaal and Bradbury 1987). When using this mating tactic, males do not defend territories, but attempt to follow herds of females which they defend against other males. Furthermore, an increasing number of papers has reported lekking in the fallow deer (Pemberton and Balmford 1987, Schaal and Bradbury 1987, Clutton-Brock et al. 1988, Clutton-Brock 1989, Appolonio et al. 1989a,b, 1990). We can thus see how fallow deer exhibit most of the different mating systems ever reported in ungulates, from nonterritorial harem defense, through dispersed territories, to highly localized mating centers as on leks.

Is it possible to understand the ecological background of this variation? Langbein and Thirgood (1989) were able to monitor the mating system of fallow deer in nine enclosed deer parks and four wild populations in Britain. Their analysis showed that the most important factor in explaining increasing territoriality was the density of deer (fig. 8.1). In low-density populations, harem defense prevailed. As density increased, males started to defend territories and, in the densest populations, males displayed on leks. Langbein and Thirgood list some other important factors such as habitat structure and tree cover, suggesting that leks are favored in relatively open habitats. Also, if the absolute number of deer drops under a critical number, leks did not seem to be able to exist. An additional important finding was that in a few sites, the density changed and the mating system also changed as expected. These populations largely consisted of the same individuals, indicating that mating system flexibility can occur within the same individuals.

Studies of lekking fallow deer populations in Italy suggest that just looking at local densities may be slightly deceptive. The Italian populations

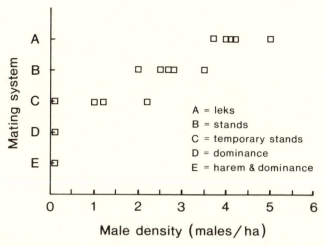

Figure 8.1 Mating system in relation to male density in fallow deer. (Modified after Langbein and Thirgood 1989)

studied by Appolonio and co-workers have moderate densities, yet lekking is the most common mating system (Appolonio et al. 1989a,b, 1990). Appolonio suggests that the environmental diversity in his study areas creates high local densities of females at certain sites of the study area (Appolonio 1989). Thus males may assemble on leks at places where the chances of encountering many females are high. This explanation resembles the hotspot hypothesis (see section 7.3; Bradbury and Gibson 1983, Bradbury et al. 1986). Yet, even though other factors apart from density may be of importance in some populations, the effects of density are impressive.

What does this mean? Density could be seen as proportional to the encounter rate of conspecifics. Thus, the economics of different mating tactics such as lekking, territory defense, and harem defense probably differ at different densities. It may be worthwhile to defend harems at low density, whereas the benefits of harem defense decline as density and the encounter rate with other males increase. At very high densities the economics of territory defense could break down, and the stable solution may be to defend a tiny court on a lek.

This shows that males can switch their behavior according to changes presumably related to intruder pressure from other males and economic defense of females.

SOME OTHER EXAMPLES

Mating system flexibility seems to be particularly common among ungulates (see chapter 2) but is also reported in other groups (Lott 1991). In bullfrogs the most common mating system is resource-defense polygyny, but leks also occur (see section 2.5; Emlen 1968, 1976, Howard 1978a,b, 1983, 1984, Ryan 1980b). In parrotfish *Scarus vetula*, lek mating and harem polygyny have been reported in the same population (Clavijo 1983). In birds, the mating system is sometimes flexible even within populations either because individuals differ (Davies 1985, 1986, Davies and Houston 1986) or the environment changes (Lank and Smith 1987, Gibson and Bradbury 1987, Pruett-Jones 1988, Cartar and Lyon 1988). Variation in individual quality and environmental factors lead to the evolution of alternative reproductive behaviors, which is the topic of the next section.

8.3 Intra-Individual Variability and Alternative Reproductive Behaviors

As pointed out by Parker (1982), one of the basic questions of evolutionary biology is the origin and maintenance of individual variation. Phenotypic variance, V_P, can be partitioned into variation due to genetic factors, V_G, variation due to the environment, V_E, such that $V_P = V_G + V_E$ (Falconer 1981). If there is no genetic variation (more specifically, additive genetic variance), there will be no evolutionary response to selection. However, even where selection is stabilizing and may act to reduce V_G, there can still be additive genetic variance due to factors such as recurrent mutation, linkage, and other factors that act against fixation of any given allele on any given locus. An obvious explanation for genetic variability is of course that alternative alleles may have equal fitness. In behavioral ecology, where the genetic system under study often is unknown, equal fitness is analyzed using ESS models. In ESS models, alternative ways of solving animal behavior patterns, called strategies, are assumed to have a genetic origin and to be able to be changed by selection. A strategy I is said to be an ESS if no other conceivable alternative by a rare mutant J is able to do better in terms of fitness gains relative to I.

Variation in a strategy can come about in two ways (Parker 1978a, 1982, Maynard Smith 1983). First, a strategy can be a pure ESS, but conditions may vary. Thus in a set of conditions $A,B,C \ldots N$ the ESS, when in A is to play x_A, when in B to play x_B, and so on. Here the payoffs may differ between any two strategies, for example x_A and x_B. Such an ESS has also been called *conditional ESS* or "best of a bad job" (Dawkins 1980). Second, there can be a mixed ESS in which conditions are fixed. Here the ESS is: when in A, play x_1 with the probability p_1 and play x_2 with p_2. In reality both varying conditions and mixed ESS may occur simultaneously, but even so, payoffs will differ between but not within conditions.

A condition can be a habitat or a phenotype given by ontogeny, say by size or physical condition. Parker (1982, 1984) has analyzed this problem,

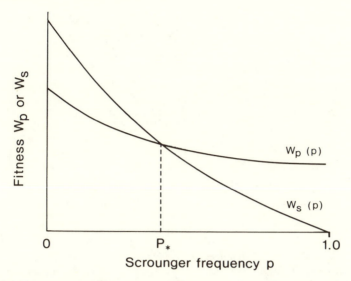

Figure 8.2 Fitness in relation to scrounger frequency in the producer-scrounger model. (After Barnard and Sibly 1981)

which is referred to as a *phenotype-limited strategy*. A phenotype-limited strategy is an ESS when there is no alternative strategy that is better when played by an individual of the same phenotype.

It is useful to start the discussion of phenotype-limited ESS with a simple model of producers and scroungers (Barnard and Sibly 1981, Parker 1984b). In producer-scrounger systems, one strategy is to search, detect, and exploit resources (presumably at some cost); these are the producers P. An alternative strategy is to avoid searching and exploit the investment of P strategists; these are the scroungers S.

In producer-scrounger systems the fitness of S is frequency dependent. In stable solutions of this game, the fitness of S, Ws, must be higher than the fitness of P, Wp, when S is rare, but lower when S is common. A graphic solution is given in figure 8.2. In this case there is a mixed ESS, and the ESS frequency of S is p_*, where both strategies have equal payoffs.

This model can be extended, leading to the model outlined below—Parker's continuous phenotype model in alternative strategy competition (Parker 1982; see also Parker 1984). Other examples of phenotype-limited ESS models also proposed by Parker are outlined in sections 3.5 and 9.2.

Let strategy X be a territory-defending strategy in which competitive ability increases with size s, and let Y be a parasitic sneak strategy in which size is irrelevant. If there is perfect phenotype limitation, selection will favor a strategy switch at threshold size T. The ESS is to play Y below T and X above T (fig. 8.3).

This model is a producer-scrounger model where the fitness of individu-

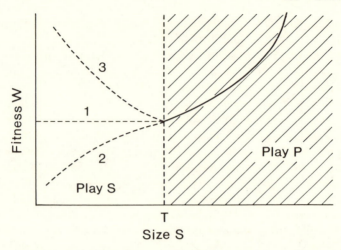

Figure 8.3 Fitness in relation to size in a hypothetical population in which there is a competitive advantage to large size in strategy *P* but not in strategy *S*. The ESS is to play *S* when size < *T* and *P* when size > *T*. (Modified after Parker 1982)

als that play *X* (territory defense) is dependent on competitive ability. When continuously varying phenotypes that affect individual payoff were included, the ESS was a pure strategy: when small, play *Y*; and when large, play *X*. In the pure producer-scrounger model there was a mixed ESS.

Phenotype-limited models seem to have a wide range of relevance to real biological systems. Size asymmetries in reproductive behaviors are reported in a number of insects (Thornhill and Alcock 1983). In vertebrates, life-history switches in which young (and often small) individuals employ different mating tactics as compared to old (and larger) ones are sometimes found (see references in Parker 1984a). Larger common toads are more likely to be found calling to attract females than smaller ones, possibly because larger individuals can attract females (passively) over larger distances. Small individuals are instead more likely to adopt the alternative mating strategy of active searching (Höglund and Robertson 1988; see also Arak 1983b for an example from natterjack toads). In lekking birds, phenotype-limited alternative mating strategies have been suggested in the great snipe (Höglund and Robertson 1990a) and may have been taken to an extreme case in the ruff (van Rhijn 1991, Höglund et al. 1993).

In ruffs there are two main strategies: (1) territory defense in which males invest in territory maintenance at a lek where they mate with visiting females (such males are called *independent males* [*I*]) (Hogan-Warburg 1966, van Rhijn 1973, 1983); and (2) the alternative mating strategy, called *satellite* (*S*). Such males do not defend territories and visit the leks most commonly with females. While on the leks, they associate with *I* males and gain access to females most commonly when *I* males are involved in fighting off other males. To complicate things further, *I* males sometimes fail to establish a lek territory and display intermittently on the edge of leks. *I* males that maintain territories are called *resident* (*R*) and intermittent males are called *marginal* (*M*). Resident-marginal male strategies appear to be commonplace in lek-breeding animals.

Behavioral differences are correlated with differences in plumage color and size in ruffs (Hogan-Warburg 1966, van Rhijn 1973). *I* males basically have a dark plumage and almost invariably dark head tufts, whereas *S* males are found mainly in light or all-white plumage. Furthermore, there is a strong sexual dimorphism in both size and plumage. Males are larger and heavier than females and during the mating season possess elongated feathers on the neck and head, creating the ruff and head tufts that females lack (van Rhijn 1991). Also, the feathers around the base of the bill can be dropped, and males can have facial wattles that vary in color from green over yellow to red (van Rhijn 1991). Leg color is also variable: young males have mostly dark and green legs and older males have red legs (pers. obs.). Plumage color and hence also behavioral strategy are also correlated with size differences. Darker males are larger than browner ones which are larger than white males (Höglund and Lundberg 1989). *I* and *S* strategies seem to be fixed for life and do not change (Andersen 1948, van Rhijn 1991; F. Widemo, R. M. Montgomerie, and J. Höglund, unpubl.; D. B. Lank, pers. comm.). This is in contrast to *M* and *R* strategies, which can change during the course of a season and with age (van Rhijn 1991). In most populations studied, the ratio of *I* to *S* males is about 3:1 (van Rhijn 1991).

The mating success of a given *I* male to a large extent seems to be determined by competitive ability. Within leks, the most dominant males seem to get a disproportionate share of the matings (F. Widemo, R. M. Montgomerie, and J. Höglund, unpubl.), and between leks, larger leks have a higher average *I* male success. These patterns seem to be less pronounced among *S* males (Höglund et al. 1993).

Comparing the annual success of *I* and *S* males, the proportion of satellite males in the population is 20%, but only 12% of the copulations are performed by satellite males (Höglund et al. 1993, F. Widemo, R. M. Montgomerie, and J. Höglund, unpubl.). However, this estimate is based on observations of matings on leks and may be changed if males also mate elsewhere (we have no evidence of this, however). If *I* males annually reproduce at a higher rate than *S* males, this difference may be offset by *S* males being able to reproduce during more breeding seasons. Further research is needed, however, to resolve if this is the case. Another possibility is that *S* males have an unusually poor reproductive rate in our population and that the stable *I* to *S* ratio is maintained by immigration of *S* males.

Can this bewildering variability in ruffs be understood? It seems as if male ruffs play at least three different games. The first game concerns the stable frequency of I to S males. If the frequency of I to S is a mixed ESS, the lifetime fitness of I males $W(I)$ must equal that of S males $W(S)$. Present evidence in a local population suggests that $W(I) > W(S)$, which means that additional compensating factors have to be invoked or our estimates are wrong.

The second game concerns the optimal investment in territory defense given I. This is a phenotype-limited ESS that is similar to the phenotype-limited ornament investment game outlined in section 3.5 (Parker 1982, 1983). When applying this game to ruffs, we need to have estimates of the distribution of competitive abilities $K(N)$, which is not straightforward in this species.

The third game is also a phenotype-limited ESS: given R and S, what is the best lek to visit, respectively? As outlined in sections 7.4 and 9.2, I males seem to fit the predictions of an ideal free distribution of unequal competitors. Fitting with this model, larger leks were visited by more mating females, male quality might be higher in such leks (indirect evidence from higher costs in such), and the average male fitness of I males increased with lek size (Höglund et al. 1993). It is more problematic to fit the distribution of S males. A possibility is that S males also differ in quality but face a different trade-off compared to I males. I males have to trade the increasing mating benefits on larger leks against the increasing costs in territory defense according to their ability to defend a mating territory against intruders and parasitism by S males. I males are thus likely to face energetic limitations. S males also gain by the increased mating benefits at larger leks; however, their costs at such sites are probably due to the limited access admitted by the high-quality I males present there.

In the great snipe it is possible that a game similar to the kleptoparasitism game exists. Male success in this species is, like in many other lekking birds, highly skewed toward a few males and seems mainly to be explained by three factors: male display rate in which males that show high rates attract and mate with more females; dominance within lek territories; and central territorial position on the lek (Höglund and Lundberg 1987, Höglund et al. 1990a, Höglund and Robertson 1990a). An additional factor that may be important is the size of the white tail spots, which is correlated with age (Höglund et al. 1990a). The effect of territorial position could be explained as an effect of less attractive/less dominant males surrounding attractive and dominant ones (section 4.2; Höglund and Robertson 1990a).

Thus it appears as if in this species less attractive birds parasitize the attractive ones, which leads to a resemblance to the kleptoparasitism model. The game could be stated as follows. Given quality/attractiveness, what is best: join a given lek or leave? Again it is hard to estimate the range of competitive abilities in real birds to make a quantitative fit with the model. Two qualitative predictions are possible, however. First, given high quality, it is easy to see that it would pay such males to move if females are attracted to the lek only by the quality of the best males (pure hotshot effect). If female input rates, on the other hand, are determined by other factors such as habitat limitation or number of males present, it may pay to stay. Little is known about what determines female input rates in great snipe. Second, given low quality, the options are to stay and parasitize or leave and go to a site with a different input rate. Here the ESS is if $W_{parasite}$ > W_{leave}, males should stay (Waltz 1982). Given the nature of great snipe leks where many males achieve close to zero success and behave as if they are mainly interested in a position close to an attractive male it seems as if $W_{parasite}$ indeed is higher than W_{leave}.

8.4 Summary

Intraspecific variation in mating systems is probably a common phenomenon. Here we have reviewed a few cases with relevance to leks, in particular the case of the fallow deer. Understanding the cause of intraspecific mating system variation will most certainly help to reveal why leks exist. More intraspecific comparisons are needed.

Mating systems may vary within species for two reasons: (1) Populations of the same species may vary because the environment is different. This may be because of differences in for example, population density, vegetation structure, resource distributions, etc. (2) Individuals within the same population may vary. Again this may ultimately be attributed to a range of extrinsic factors.

Individual variation may be analyzed with ESS modeling. If conditions are fixed, variation can come about because two strategies are in a mixed ESS, that is, they have equal evolutionary payoffs. A possible case of a mixed-ESS in lekking animals is the indpendent-satellite strategies played by male ruffs.

If conditions vary and strategy A is the best to play given the environment X_A, the strategy is a conditional ESS. If strategies are conditional upon phenotype, we usually talk about phenotype-limited strategies. An example of a possible phenotype-limited strategy in lekking animals is in the great snipe, where an unattractive male will parasitize close to an attractive male as an alternative to self-advertisement.

9   Game Theory
       Models of Leks

9.1 Introduction

Neo-Darwinian doctrine states that all individuals are selected to behave in order to maximize their reproductive success. In doing so they have to compete with other members of the same species and overcome the problems of finding food and avoiding enemies. In many species, including the lekking ones, much of the variance in male lifetime reproductive success comes from variation in the number of females inseminated. When trying to mate with as many females as possible, males face problems of mate acquisition. More specifically, they have to solve problems of how, when, and where to search for mates. Thus an important aspect of understanding leks is to understand the selective forces caused by sexual competition, which are generated either by female mate preferences or male-male combat. Female spacing behavior will also affect male mating success, mainly in affecting where males compete for females. In this chapter we attempt to analyze the question of where to compete.

The searching behavior of males can be thought of as strategic "decisions." These are not decisions in the general sense, but behaviors or other traits shaped by natural selection. Traits that are relatively less successful have been eliminated over evolutionary time. In "deciding" what to do, it is important to take into account what other competitors are doing. With game theory, biologists can analyze how individuals should decide given what others are doing. In evolutionary terms this is the same as asking which strategies can resist invasion by other strategies and be stable over evolutionary time (an ESS) (Maynard Smith and Price 1973).

9.2 Ideal Free Theory

A concept used in analyzing animal distributions is the ideal free distribution theory (Fretwell and Lucas 1970, Fretwell 1972, Milinski and Parker 1991). Such models ask how individuals should distribute themselves to maximize their intake rate of resource items. Resource items could be thought of as food or breeding sites and, in leks, as the number of mates.

Encounter Sites

In lek mating systems, matings occur in patches. Given the above framework, the clustering of males could be understood in terms of clustering of females (fig. 9.1). There are at least two possible reasons why females should be initially clustered, and in the following sections we describe a range of models with different assumptions. Possible reasons for female clustering are (1) resource limitation and (2) overlapping home ranges that result in hotspots. In any given population, both of these reasons can be important. In addition there are two more explanations that rely on female behavior, but not on resources or female clustering, and which may explain why males should aggregate: (1) the nearest-neighbor distances are unstable (kleptoparasitism or the hotshot mechanism; section 9.2); and (2) the nearest-neighbour distances are stable (black hole theory, section 9.2). We will first turn to models that assume that females are clustered initially.

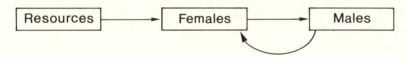

Figure 9.1 Presumed relationships between the distribution of resources, females, and males in non-role-reversed species. The feedback between males and females may not always occur.

Fixed Sites, Equal Competitors

As a starting point for game models of leks it is useful to start with the simplest possible case, even if it does not represent a lek. This is to assume that there are a fixed number of sites where matings can take place. Females are clustered at such sites because resources essential for females are clustered. Such models are equivalent to models of the ideal free distribution of individuals competing for food in patches or oviposition sites (Fretwell and Lucas 1970, Fretwell 1972, Sutherland 1983, Courtney and Parker 1985). An individual's gain rate W_i in a site i depends on the number of competitors n_i present in the same site. For the distribution of males in all sites to be an ESS, all individuals must achieve the same payoffs in all sites. This means that

$$W_i(n_i) = c \text{ (constant)}$$

for all sites $i, j, k \dots n$.

In general, an individual's gain rate (for leks this is equal to male fertilization rate) is given by

$$W_i(n_i) = Q_i n_i^{-m} = c, \text{ for all } i, j, k \ldots n, \tag{1}$$

where Q is the input rate of females to site i, and m $(0 < m < 1)$ is a constant that describes the effect of interference (Hassel and Varley 1969).

In continuous input systems, all resource items are consumed, and for such systems m must equal 1 (fig. 9.2; Sutherland and Parker 1992). Females on leks can be thought of as being supplied continuously, and therefore

$$W_i(n_i) = Q_i / n_i, \tag{2}$$

which gives the input matching rule,

$$n_i = Q_i / c, \tag{3}$$

that the number of males on a given lek should match the input rate of females (Parker 1978a). This simple model thus tells us that if the ideal free distribution is correct, the number of males in a lek should be proportional to the number of females.

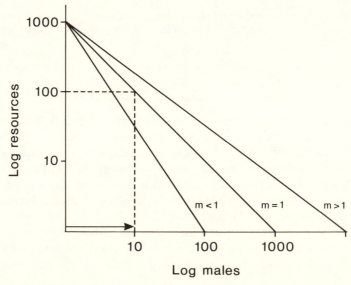

Figure 9.2 The relationship between log resources and log number of competitors in a site. In continuous input systems, resources are consumed immediately as they appear, like females on a lek. Therefore a tenfold increase in the number of males present must devalue the site tenfold and $m = 1$. Values of $m < 1$ are possible if females are repelled from larger lek sites, for example by increased male disruption. Values of $m > 1$ indicate cooperation. Neither of these possibilities is likely to have a great effect on leks.

Hotspots, Males Equal

The above model has obvious limitations when applied to leks. One assumption that may be oversimplified is that leks are discrete with no female movement between them. Bradbury and Gibson (1983) noted that a typical feature of many vertebrate lekking species is the mobility of females. This means that a female, by mating on a lek, not only devaluates this site but also neighboring sites within her home range. The hotspot model is an attempt to account for such devaluation.

The modeling of such a system may be solved by the simulations described in section 7.3. An alternative approach is to use the binomial distribution in determining the number of females on each lek (Sutherland, unpub.). Sutherland considered female territories positioned randomly over a random number of lek sites. The number of overlapping female home ranges at each site expresses the amount of female traffic, and the probability that a given female home-range covers a given lek site is given by home-range size divided by the total available area.

The relationship between the distribution of males and females is the same as given by equation (3). Female home-range size affects lek size: when female home ranges are very large, the number of overlapping home ranges in each site approaches the total number of female home ranges. In the extreme case of female home ranges covering all the available habitat, all home ranges cover all sites. Male clustering is highest at intermediate home-range size, and when home ranges are small all males will stay with one female (see also Bradbury et al. 1986).

Differences in Competitive Ability

Another obvious limitation of the basic model (eq. 1) is the assumption that all males are equal in competitive ability. Individual differences can be modeled using the same framework as above (Sutherland and Parker 1985, Parker and Sutherland 1986). The fertilization rate of an individual male in this case is given by

$$W_i = Q_i n_i^{-mR},$$ (4)

where Q and m are the same as in equation (1), and R is the competitive ability of a given individual, K_j, divided by the mean competitive ability for the habitat K. The distribution of each phenotype can then be determined assuming that each individual mates in the site in which its rate of fertilization is highest. In simulations this can be achieved by distributing all phenotypes equally and calculating their success rates. In following iterations, the number of each phenotype is directly proportional to their success. This can be thought of as individuals moving between sites while assessing their own abilities and future prospects. At the ESS distributions, the success rate will be the same for all individuals of a given phenotype although this will differ between phenotypes.

The solutions to the unequal competitor game are complex, and there are many equilibrium solutions (Sutherland and Parker 1985, Parker and Sutherland 1986). Both unstable and stable equilibria exist and only the stable are possible as ESS's. A major finding of such ESS distributions is that high-quality individuals will be found in richer sites and low quality individuals in poor sites (fig. 9.3; Sutherland and Parker 1992).

The effects of the range in competitive abilities can be seen in figure 9.3. For all ranges of competitive abilities, individuals are forced to poorer sites compared to the ideal free case. This effect is more pronounced as the range of phenotypes increases. Allowing for differences in competitive ability will thus decrease the clustering of males and cause a more even spread among the lek sites.

The same results are also found when lek sites are not fixed (hotspot settlement) and males differ in competitive ability. High-quality males are found in richer sites, low-quality males in poorer ones, where male clustering is also reduced (Sutherland, unpublished).

Bradbury et al. (1986) incorporated the effects of despotism in some of their simulations of hotspots. Despotism has the effect that subsequent settlers achieve a disproportionately smaller share of the available resources. Mathematically this is similar to allowing differences in competitive ability. Each female contacted was worth $P(R)$ copulations to the Rth settler within her range, as

$$P(R) = \frac{R^{-1.8}}{\displaystyle\sum_{J=1}^{R} J^{-1.8}}$$

where the exponent -1.8 was chosen to mimic observed values of male mating skew. This approach has the same result as the unequal competitor approach: by allowing despotism, male clustering is reduced.

In summary, by incorporating individual differences and despotism, the models give three main predictions: (1) there is a correlation between lek size and male quality at least for some, and probably for the most realistic, parameter values; (2) average male success is higher in larger leks; and (3) differences and male clumping reduces male clustering by forcing poor-quality males to poorer sites.

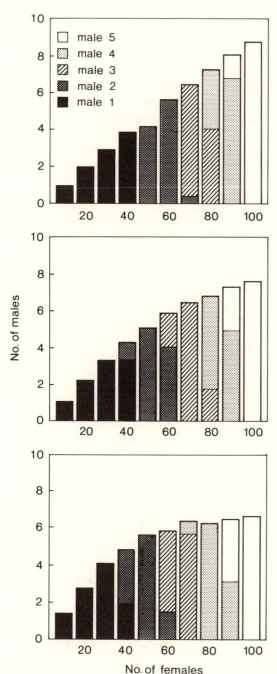

Figure 9.3 Simulated number of males in ten sites in relation to female density. Each graph represents three different ranges in male competitive abilities, from small (*top*), intermediate (*center*), to large (*bottom*) differences. The distributions of five phenotypes ranked from the worst (male 1) to the best (male 5) phenotype are given as different shades in the columns.

Kleptoparasitism

The basic model (equations 1–3) can be extended to study effects of klepto-parasitism (Parker and Sutherland 1986). This is when dominant individuals steal resources from subordinates. Within a patch, each individual has an equal probability of finding a resource item but is less likely to keep it if surrounded by dominant individuals. This is thus a model of the hotshot mechanism where dominant individuals are surrounded by subordinates that parasitize the attractiveness of the dominant.

For the continuous input case (which is relevant to the lek situation), the fertilization rate of a given individual in lek i can be modeled as

$$W_i(n) = Q_i(1 + G(n_s) - L(n_d))/n_i, \qquad (5)$$

where 1 is the number of females attracted by self, $G(n_s)$ is gains from subordinates, $L(n_d)$ is losses to dominants, and n_i is the total number of competitors in i. In this model total loss to all individuals in i must equal total gains, and the interactions among males do not affect the average intake rate. Instead, interactions between males result in a reallocation of matings.

In general, there are no stable solutions to this game that contrast with the ESS's found for the previous models (equations 2 and 4) (Parker and Sutherland 1986). In simulations with three phenotypes and ten sites, Parker and Sutherland found that the most dominant phenotypes tended to be confined to the best site, intermediate to the next best site, while the least dominant tended to fluctuate among all sites. If such a system would be sampled occasionally and then averaged, there would be an overall correlation between phenotypic quality and site richness—a result that echoes the one obtained in the unequal competitor case. However, except for the most dominant phenotype, there will be much variation, and low-ranking subordinates will be found in all types of leks. Thus this kind of model predicts continual movements (especially of low-ranking individuals) between lek sites, and the fertilization rate of a given phenotype will not be the same across lek sites. It remains to be shown if such properties can be found for lek species in the wild.

Female Preference

In previous models females have been assumed to be indiscriminate about mating sites. Assuming instead that females discriminate among leks, it is slightly more complex to determine the number of females in a given lek (Sutherland, unpubl.). One approach to find the number of females in a given site is to assume that the number of females is a simple function of the number of males present, such as

$$Q_i = c(\alpha\, n_i - \text{ß}\, n_i^2),\tag{6}$$

where c is a normalizing constant and α and ß are constants that describe the precise manner in which the distribution of females is determined by the number of males, n_i. Varying α will vary the benefits of mating in larger leks, whereas varying β will affect the costs.

Depending on the values of the costs and benefits, which are proportional to the constants a and b, males will either congregate in a few sites or stay dispersed (fig. 9.4). Some combinations of values of α and β are obviously unrealistic. It is impossible for the costs to be higher than the benefits, or else females would not mate at all. However, at least for some parameter values, female preference will cause enhanced male aggregation and thus be a prime factor in lek evolution. Note that the benefit for females mating in larger leks need not be genetic even though it could be so. A possible reason for why females should prefer to mate in larger leks is the correlation between lek size and average male quality (eqs. 2, 4, and 5) (Alatalo et al. 1992, Höglund et al. 1993; see section 7.7).

Bradbury (1981) suggested that female preference for larger leks would cause males to cluster until there was a single lek per population or per female home range. Leks should thus be spaced an average female home range diameter apart and each female should visit only one lek. This result depends on a number of assumptions: among initial leks, one is always larger; females always detect lek size without error; and females do not move out of their home ranges to mate.

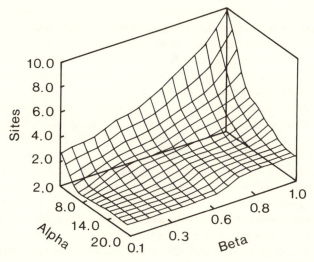

Figure 9.4 Simulated number of sites when females are allowed to move in relation to the parameters α and β. Original number of sites is 10, α scales the benefits and β the costs of females mating in larger leks.

Gibson et al. (1990) produced a model without the assumption that females have a restricted home range. Instead, in their model, female choice of lek size depends on the travel costs from the nest site to a given lek, the number or quality of the males present, and competition among females for access to males. Males and females were initially settled randomly on a random square within a grid and were then allowed to move, given the rules below, until no further movement was observed.

Males assessed squares on the basis of fertilization rates and then moved to the best square. In this simulation it was possible to run separate subprograms for all individuals so that at each iteration males assessed squares on the basis of the current distribution of females (fertilization), and all males then moved simultaneously.

For females, overall desirability D_i of site i was computed as

$$D_i = k_1(-d_i) + k_2(n_i) + k_3(f_i/n_i),$$

where d_i is travel costs from the nest to the lek site, n_i is the number of males present, and f_i/n_i is the number of females per male present. The relative weights of these variables were adjusted by allocating each k_i so that $k_1 + k_2 + k_3 = 1$. This model could be slightly adjusted to study the effects of male quality instead of male numbers. The term n_i is simply replaced by "the quality of best male present" that could vary from 0 to 1 drawn from a uniform distribution, and d_i was rescaled to make the range of values comparable. Females examined all squares and stayed in the most desirable one.

Each of the three assumptions (search costs related to distance, search costs related to area, and female choice of best-quality males) could generate a broad range of outcomes from single leks to overdispersed males depending on parameter values. If the number (or quality) of males was given high value a single lek was the most likely outcome. This suggests that when preferences are sufficiently strong, males become clustered. If travel costs and the benefit of many males were given intermediate values, many small leks were found. Thus increased travel costs could offset the more-males advantage and stop further clustering. Females would still visit several leks, contrary to Bradbury (1981). If travel costs were given a very high weight, males became widely separated and the final male distribution mirrored the initial female one. High values of the female to male ratio and low weights for the number of males on the lek resulted in overdispersion, and when the female to male ratio was given low weight (females minimizing the competition for males) males were found in separate squares.

These simulations show the interactive nature of female preference on the dispersion of both sexes. Since males disperse in relation to females, and females, at least theoretically, can be made to be influenced by male dispersion, the outcome becomes an interactive game in which the solution can be very complex. The relative importance of female preference for larger leks or leks with better males depends on the relative weights of factors such as travel costs and competition among females.

A different approach to the question of why males should cluster than the one reviewed above is the black hole model (Clutton-Brock et al. 1992, Stillman et al. 1993). This model rests on the assumption that females are harassed by males when entering their territories and hence move in a random direction until they eventually mate. The model seeks the optimal nearest-neighbor distance between males given that females behave in this way. The model was explicitly based on observations of female ungulates in which there is some evidence that females are harassed by young males, that such matings are dangerous, and that females are also frequently forced to move from territorial males because of disruptions (Clutton-Brock et al. 1992). At least theoretically, female movement could also be caused by other factors, such as sampling sequences, when females have preferences for particular males.

The optimal nearest-neighbor distances can be shown to be quite small such that the males are clustered (fig. 9.5). This result can be understood by an analogy with a pinball game. Think of the habitat in which females move as the box of the pinball game, the females as the balls that enter the pinball machine, and male territories as the bumpers on which the females bounce. If bumpers are spaced, the ball will be bounced off in any random direction and is not likely to return, whereas if bumpers are clustered the ball tends to be sucked into this black hole, and the probability that a given bumper within a cluster will be hit again is increased. Given that females mate after a given time, they are more likely to mate while in a cluster than with any male that is far from others. The females in such a system thus do not actively choose to mate on a lek, but end up in leks as a result of the process of being evicted from territories.

While these assumptions appear valid in ungulate mating systems and in a marine iguana (F. Trillmich, pers. comm.), any process that resembles bouncing due to harassment is not obvious in bird and insect leks. In ungulates and iguanas, females may be forced onto the lek (Clutton-Brock et al. 1992), whereas in many birds and insects it seems as if females arrive unharassed to the lek. Furthermore, in birds the lek is not a safe place with regard to harassment. On the contrary, on bird leks harassment is a common phenomenon (Trail and Koutnik 1986).

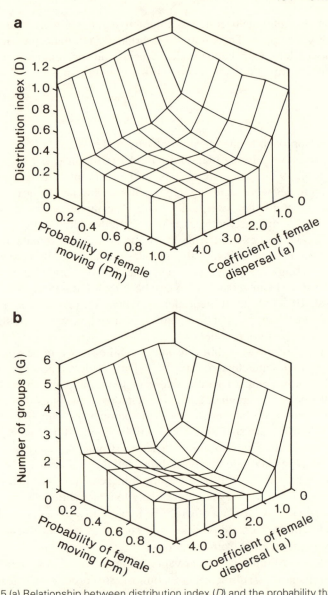

Figure 9.5 (a) Relationship between distribution index (*D*) and the probability that females move when entering male territories (*Pm*) and a coefficient of female dispersal (*a*) in the black hole model. *D* approaches zero as all males are clustered and increases as males are more spaced. *a* increases with female range. (b) The number of male groups (leks) in relation to *Pm* and *a*. (After Stillman et al. 1993)

9.3 Summary

The models presented in this chapter all vary in complexity and in their underlying assumptions. They should not be taken too literally. Even a very simple model that obviously violates the real world is useful in making the assumptions explicit and in helping us to understand the outcome of the specified system.

The basic ideal free model (eqs. 1–3) explains male clustering simply by the fact that there are more males than sites in the population. This situation is very unlikely to be the entire explanation for male aggregation in most of the species discussed in this book. A system that may fit the basic model is that of the non-lekking yellow dung fly (Parker 1970), which mates on patches of dung. However, even in this species, males are most certainly not of equal competitive ability (even if the differences are small), and thus the model where males were allowed to differ in competitive ability (eq. 4) may show a closer fit. A species in which this model shows a qualitative fit is the ruff, in which males that attempt to defend territories seem to distribute themselves according to competitive ability (Sutherland et al., in prep.).

Including kleptoparasitism (eq. 5) in the theoretical model makes the system unstable, and especially poor males are predicted to move frequently. A possible case of kleptoparasitism in great snipe was described by Höglund and Robertson (1990b). In this species there was evidence that poor males settle around successful ones, which is a basic prediction of the hotshot model. In addition, poor males are more likely to change leks. Unless there are habitat limitations, the lek aggregations are theoretically predicted to move frequently. Habitat restrictions may be common in the wild, as in the case of many amphibians that must breed in water (e.g., Arak 1983b). However, when habitat limitations are included in theoretic assumptions, the difference from the basic model (eqs. 1–3) where aggregation of males is explained by a shortage of sites becomes less clear.

In all the equations (1–5), females were assumed to be fixed to a given site and sites were assumed to vary in female numbers. The variation in female numbers is assumed to be due to differences among sites in the distribution of resources that in the case of lek-breeding animals cannot be defended by territorial males. The assumption of fixed females is clearly not true for most lekking species. The hotspot model is an attempt to overcome this problem. In this model, the habitat is uniform and all female clustering is explained by their overlapping home ranges. We thus have two extremes: (1) the basic model (eqs. 1–3), where the aggregation of females is determined by resources, and (2) the hotspot model, where resources have no effect. In real life both overlapping home ranges and a

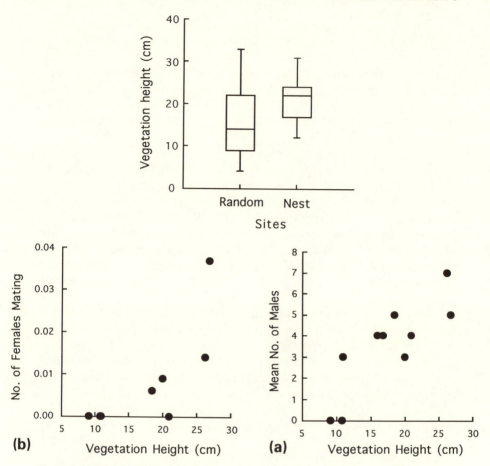

Figure 9.6 (*Top*) Vegetation height at nests sites of ruffs and random sites. Boxes show 5th percentile (lower bar), 25th percentile (bottom of the box), median (line through box), 75th percentile (top of box), and 95th percentile (upper bar). (*Left*) The number of females mating on seven leks in 1992 (corrected for observation time) in relation to vegetation height. (*Right*) The mean number of males present on ten leks in 1992 in relation to vegetation height.

clustered resource distribution are likely to have an effect. A recent study of female ruffs, a species in which females are known to show extensive movement, showed that both female and male numbers at various leks could to some extent be explained by a resource. Both the number of females and the number of males at leks in a limited study area were correlated to the height of the vegetation surrounding the leks (Sutherland et al., in prep.). Evidence suggests that females prefer to breed in high vegetation. Thus, in this species, the link between resources (linked to high vegetation), female numbers, and male numbers was established (fig. 9.6).

In a study area in Costa Rica, Westcott (in prep.) showed that leks of four species tended to be located in the same place year after year. Three of these species—a tyrant flycatcher and two species of manakins—are frugivores and had overlapping diets. They were more similar in lek location and changing numbers than the fourth species, a trap-lining nectar-feeding hummingbird. These results suggest that a common environmental factor sets limits to where leks can be located, in this case the topography of the landscape. However, the fact that the frugivores were more similar also suggests that food resources to some extent also determine lek location and the change in number of males on the leks in different years. These results again suggest a role for resource distribution.

Female preference for aggregated males can (e.g., eq. 6), depending on the parameter values, cause male aggregation. Female choice, however, is not a necessary condition for male aggregation and can under some circumstances even cause overdispersion. An interesting possibility is a link between the model of unequal competitors (eq. 4) and female preference. Since equation 4 predicts that larger leks should contain better than average competitors, there is a rationale for females preferring to mate in such sites. Female preference for mating in larger leks has been suggested in black grouse (Kruijt et al. 1972, Alatalo et al. 1992) and ruffs (Lank and Smith 1992, Höglund et al. 1993).

The models based on male harassment (black holes) differ from other models of lek evolution in that females are not fixed to a site by resources, do not necessarily have overlapping home ranges, nor do they prefer to mate in clusters. All females do is avoid sexual harassment, which they do by visiting the leks because these are the safest places to be. By constantly avoiding males, females can cause males to aggregate, because females are more easily retained in male aggregations. Recent studies of fallow deer (Stillman et al. 1993, Clutton-Brock et al. 1992) and marine iguanas (F. Trillmich, pers. comm.) suggest the feasibility of this process. It is our guess that male aggregation as a cause of harassment of females is most often found in relatively immobile species.

PART IV   Conclusions

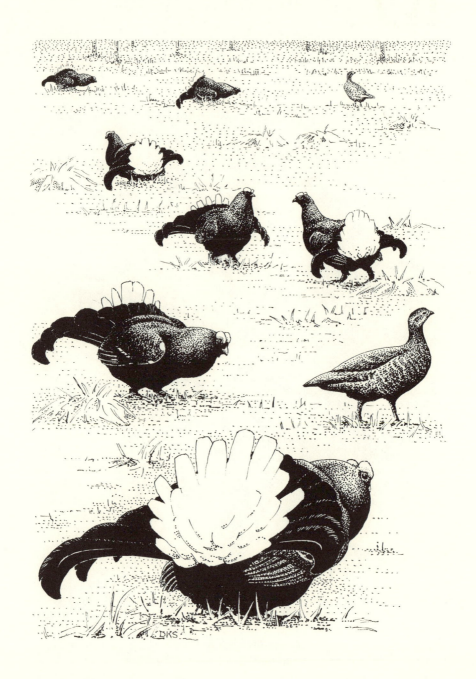

10 Concluding Remarks and Prospects for Future Studies

10.1 General Conclusions

In any imaginary species, what are the factors that will lead it to mate on leks, and how do these factors relate to the models we have outlined? The problem in all empirical biology is that in explaining an observed phenomenon, in this case a mating aggregation, we have to make inferences based on our observations. Thus we have to interpret our often sketchy knowledge and make guesses about the mechanisms we think caused the population under study to lek. In theory there can be a range of mechanisms that lead to superficially similar behavior. In a theoretical model we have the advantage of specifying the mechanism and studying the outcome. It is then up to empirical researchers to decide if any given model is applicable to any given species studied in the wild. Our proposition in this book is that leks can be caused by a range of ecological factors, selective pressures, and evolutionary mechanisms. To make this argument more clear, we summarize below what we think are the most important factors causing lekking behavior.

In chapter 2 we argued that in order for leks to evolve, males need to be liberated from paternal care. This is not a sufficient condition, however, since there are many species that show no parental care that do not lek. Therefore, a further prerequisite is that males abandon resource defense or defense of females as tactics to gain matings. Instead, males seem to use a self-advertisement strategy. If they do, this need not result in leks. When males do not defend resources or females but display widely spearated, we talk about solitary display. Such mating systems are very similar to leks; they differ only in the spatial aggregation of the males. In fact, when comparing species there is a continuum from solitary display to tight classical leks. This has led some authors to include solitary display in their definition of leks (e.g., Prum, in press). From an ecological standpoint, however, the spatial aspect is very important and has led a number of people to suggest a plethora of hypotheses for why males aggregate at arenas. Below we summarize what we think are the most plausible explanations. We want to stress that explanations for why males aggregate on leks probably have much in common with explanations for other kinds of animal aggregations, such as, for example, colonial breeding (Wagner 1993).

We have suggested that if females are initially clustered by habitat limitations or hotspots, this can lead to an aggregation of males (see sections 7.5, 7.6, and 9.2). In some species, for example in dungflies, these occurrences may be all that is needed to explain the mating aggregations observed in nature. However, they are unlikely to be a sufficient explanation for the degree of clustering in truly lekking species. If females are clustered, we must answer why a single territorial male does not defend the clustered females. An explanation may be that disruption and intruder pressure will lead to males abandoning females and/or resource defense and instead opt for self-advertisement. This is the environmental potential for the polygyny argument where lekking is considered a default strategy (e.g., Deutsch 1994). If males are then forced to lek, they should display at sites where they are likely to encounter many females (hotspots).

An alternative to the above scenario is that if males are likely to lose their females before they have had the chance to mate with them and if clustered males are likely to retain females faster than males who are spaced out, males may be selected to settle close to one another (black hole theory; see sections 7.8 and 9.2). Without knowledge of the underlying mechanism, a field biologist may think that the mating aggregations of two different species (one caused by habitat limitation and one by a black hole process) are similar and may want to seek a common explanation.

A third alternative to both of the above explanations may be that in an initially dispersed advertising system, some males may be so discriminated against by female mating preferences that their best option is to join an attractive male (a hotshot; see sections 7.9 and 9.2). As we have argued, this scenario by itself is unlikely to explain why leks are commonly found at the same place. However, where there are site limitations, as for example in many anurans who breed in ponds, it may be a plausible mechanism.

Given that males have to cluster, how should they distribute themselves over the available options? This depends on the range of competitive abilities in the population. If the range is small or nonexistent, the distribution of males should be ideal free with respect to females. If, on the other hand, the range is large, males should be distributed according to an ideal free distribution of unequal competitors. In the former case, average male fitness should not vary with lek size, whereas in the latter it should increase (see section 9.2).

If females for some reason prefer to mate in clusters, further clustering of the males is enhanced. We have suggested that a possible reason for a female preference of clustered males is that larger leks may contain high-quality males (because of the ideal free distribution of unequal competitors). This argument thus provides a rationale for female choice among lek sites. Depending on the costs and benefits of choosing among the sites, a range of outcomes is possible. If the benefits are sufficiently strong relative to the costs, the selective pressure on the males to join even larger aggregations is enhanced.

Within the leks, males are also likely to be spatially organized according to their competitive ability and/or attractiveness. We have argued that a mechanism similar to the hotshot argument is the most likely explanation for why central males often are observed to gain more matings than peripheral males. This is because poor males may gain additional matings by the spatial spillover of matings from the better males (see sections 3.3 and 4.4).

Differential competitive abilities and attractiveness are also likely to affect the mating strategies adopted by individual males. If the ability of gaining access to females is affected by age, experience, physical condition, and other environmental factors, the mating strategies employed by different males can be explained by the evolution of phenotype-limited strategies (see section 9.2). In rare cases, alternative mating strategies can perhaps also be explained by a mixed ESS, such as the independent and satellite strategies employed by male ruffs.

We have argued that both males and females of animal species are involved in many complex interactions in which the kinds of actions employed by individuals often depend on what others are doing. The social systems of animals, including their mating systems, can thus be analyzed by using game theory. Given the ecology, demography, life histories, and social interactions of any given population, sometimes the games these animals play result in mating systems in which the advertising sex is aggregated in a certain space. If the nonadvertising sex visits these aggregations primarily to mate, such aggregations are called leks.

When analyzed by game theory, the evolution of leks is not a mystery. Instead, leks are logical outcomes of the selection that acts upon male and female behavior. However, future work is needed to assess the relative importance of parameters such as resource limitation, including the effects of predation pressure, female home-range overlap, density, possible female preferences, and other factors that could cause males to move from their initial range into a cluster. It is also important to asses how phenotype-limited differences impact both male behavior and female choice, and how differences between individuals affect their mating success and distribution between and within leks. While much knowledge about leks has been gained over the years, contemporary studies have just begun to ask questions about the parameters listed above. We can only urge future lek studies to take up this challenge.

10. 2 Future Prospects

We hope that this book will increase the research interest on lek mating systems in all kinds of animals. In this final section we will point out the main questions that need to be studied in the future. In the past much of the work has been done on birds and ungulates, but future studies of mating aggregations in other animal groups will provide us with deeper insights into the lekking phenomenon. The term "lek" has frequently been defined in quite a narrow sense, overlooking many kinds of male aggregations visited by females primarily for the purpose of mating. All of these aggregations form a reasonably well-defined phenomenon to be studied within the same theoretical framework, even if different mechanisms are often likely to have a very different impact on the evolution of such aggregations in each case.

FEMALE CHOICE

Let us look at the current situation with respect to sexual selection and female choice. This has been a main theme in studies of behavioral ecology over the past few years. Major advances in theoretical modeling (Grafen 1990a,b, Iwasa et al. 1991, Pomiankowski et al. 1991) have now settled the debate between the "good genes" and the Fisherian approach. Both are possible in theory, and it now remains for empirically oriented researchers to design proper tests between the hypotheses. Unfortunately this will not be easy, since the critical tests would require the estimation of indirect fitness consequences of female choice. A possible shortcut for empirical tests would be if theoreticians were able to provide separate predictions from the two types of models. Testing such predictions might be much easier than unraveling the main assumptions about the heritability of fitness and male attractivity.

The heritable effects on the general viability of all the offspring and/or on the aesthetic properties of male offspring are not likely to be very large. Indeed, the progress in research has almost stopped because of the difficulty in estimating such. Further complications are caused by the need to estimate the indirect fitness effects under reasonably natural conditions. For laboratory studies it will always be necessary to consider the generality of the results under natural conditions. There is a twofold reason for this: any fitness-related traits have to be treated as reaction norms to the variability in the environment, and it is the environmental conditions that determine the strength and type of selection on such traits. Furthermore, any heritability studies will require that females do not themselves choose their mates, rather it is the experimenter that should allow each female to mate with males of given attractivity. It may even be necessary to prevent females from knowing what type of male they mated with, since a recent study on peahen (Petrie and Williams 1993) indicates that females adjust their reproductive effort in relation to the quality of the male they mated with.

Not many studies will be able to achieve all of these requirements, but well-planned experiments on this question are badly needed. Some of the lekking species do provide suitable study animals for such large-scale studies. Solving these questions will be of great importance also for the under-standing of the role of genetic variability in animal populations—for in-stance with respect to disease resistance and sizes of the minimum viable populations. This knowledge is central for the conservation of endangered animal populations, and thus funding for large-scale studies is essential.

While it will be some time before we are likely to learn much about the indirect fitness effects on female choosiness, we believe that in the near future much will be learned about possible direct benefits. Particular atten-tion should be paid to estimating the effects of sperm quality on fertility and the risks of disease transmission during mating. However, since lek mating systems are highly variable, other kinds of direct benefits may be important as well, at least in some species.

Instead of emphasizing only the benefits of female choosiness, it would be very useful to study the costs of choice among females and more gener-ally the nature and variation in female choice strategies. The female point of view (Ahnesjö et al. 1993) will be necessary to set the context for any benefits that choosy females might achieve. This information will be par-ticularly valuable also for explaining why some species have leks and oth-ers don't. Further attention also has to be paid to the female-versus-male roles in the matings, and novel experimental designs are needed to test to which degree females will mate with males other than the ones that are optimal from their point of view. It is essential to acknowledge that what seems to be a male-driven, dominance-based mating system may also be a system that serves the interests of the females.

EXPERIMENTS ON LEKS

In spite of the popularity of sexual selection studies, surprisingly little has been done in the way of experimentation to determine what traits are essen-tial for mating success. Experiments carried out so far have often dealt with morphological ornaments, which may not be the most important sexually selected traits in general in lekking systems. It would be very informative to widen the scope and to include more experiments on behavioral display traits and on territoriality and spatial aspects. We have to recall that it is likely that different male traits will be important in different species. For instance, the old idea of females preferring tightly clumped central males has only seldomly been tested, and the results so far speak against this idea. However, some old and insufficiently repeated experiments on grouse sug-gest that centrality might be a cue for female choice in this group. Like-wise, more experiments are needed to resolve the importance of female copying in mate choice.

While some experiments have been done on the choice of males *within* leks, corresponding experiments *among* leks are practically nonexisting (but see Lank and Smith 1992). Manipulation of lek size or the magnitude of the long-distance display will resolve whether active female choice is influential in the species in question. Otherwise some other mechanisms such as male harassment of females may be causing the aggregations to evolve. If females prefer aggregated males, manipulations on the structure of the lek in terms of the types of males present would determine whether females prefer aggregated males per se or if aggregations are caused by the hotshot mechanism. In addition to field studies, laboratory experiments may be well suited for certain species (see Droney 1992, 1994). Some of the lekking insects might be suitable for such experiments, as exemplified by the studies of cricket aggregations.

MALE STRATEGIES

In addition to identifying male traits under sexual selection, it will also be necessary to consider the variation in individual male investment in sexual ornaments and display. While benefits in terms of increased mating success have been commonly considered, very little attention has been paid to the costs of sexual investment in males (but see Vehrencamp et al. 1989, Höglund et al. 1992b). Indeed, costs of sexual display have rarely been considered in any animals, even if understanding the optimal sexual invest-ment of each male requires the knowledge of both the benefits and the costs. In long-lived animals, where males survive over several mating sea-sons, it will also be necessary to understand how males should change their investment over their life span.

Not much work has been done on optimal investment over the life span of lekking males. Such studies will be necessary in order to understand the organization of leks and distribution of males among different lek sites. It will be of interest to see how costly different types of male ornaments and display characteristics are, and how male investment varies accordingly. These studies will be highly interesting in particular in species that have variable male strategies with lekking and non-lekking individuals, or that have several types of lekking strategies. Also, male sexual behavior may involve different types of components in any lekking species, for instance relative investment in female attraction and harassment during copulation attempts by other males.

LEK EVOLUTION

In the previous chapters we outlined the current models of lek evolution, male distribution over lekking sites, and alternative male strategies on leks. These models specify the effects of each mechanism, and it is likely that we are already aware of most of the mechanisms that are potentially important in determining male aggregation. However, many of the models and even one of the mechanisms (female harassment by males) are very new, and we can expect further modeling development in the near future.

We already emphasized the need for experiments to manipulate lek size and structure. Another new avenue for field studies would be to use comparative analyses for groups of animals that include both lekking and non-lekking species to reveal the ecological conditions that favor lekking. So far such analyses have been possible only for ungulates, birds of paradise, and grouse. No doubt similar and larger-scale comparisons using proper methods in other animal groups would be highly valuable. This requires that the relevant field data are gathered from many new species. Detailed observational and experimental studies in groups other than birds and ungulates are also needed to widen our understanding of leks in general.

10.3 A Final Word

While the emphasis has been in unraveling the traits influenced by sexual selection and on the evolution of lekking, our focus should also be directed toward understanding the variability of the organization of leks. Questions to be asked and answered include the following: What causes the variation in mating skew? What determines the level of male aggregation? How are roles of male dominance and female choice established? And what determines territoriality? Answers to such questions will also aid in learning the mechanisms of sexual selection, the benefits and costs of female choosiness, and the mechanisms of lek evolution.

Leks continue to be one of the most productive subjects for the study of sexual selection and the benefits that females might be achieving for their offspring or for their immediate fitness. While these benefits may appear too small and thus are difficult to measure, they exert a directional selection on males that has had very remarkable evolutionary consequences. Understanding the evolution of male mating aggregations in animals in general is in itself a highly interesting task for future research. Indeed, we predict a very rapid development in this area of research in the near future.

References

Ahnesjö, I., Vincent, A., Alatalo, R., Halliday, T., and Sutherland, W. J. (1993). The role of females in influencing mating patterns. *Behav. Ecol., 4*, 187–189.

Aiken, R. B. (1982). Theories of sexual difference: The sexual selection hypothesis and its antecedents, 1786–1919. *Queast. Entomol., 18*, 1–14.

Alatalo, R. V. (1981). Problems in the measurement of evenness in ecology. *Oikos, 33*, 46–54.

Alatalo, R. V., Burke, T., Höglund, J., and Lundberg, A. (In prep.). Paternity, copulation disturbance and female choice in the black grouse.

Alatalo, R. V., Glynn, C., and Lundberg, A. (1990a). Singing rate and female attraction in the pied flycatcher: An experiment. *Anim. Behav., 39*, 601–603.

Alatalo, R. V., Gustafsson, L., and Lundberg, A. (1986a). Do females prefer older males in polygynous bird species? *Am. Nat., 127*, 241–245.

Alatalo, R. V., Gustafsson, L., and Lundberg, A. (1990b). Phenotypic selection on heritable size traits: Environmental variance and genotypic response. *Am. Nat., 135*, 464–471.

Alatalo, R. V., Höglund, J., and Lundberg, A. (1988). Patterns of variation in tail ornament size in birds. *Biol. J. Linn. Soc., 34*, 363–374.

• Alatalo, R. V., Höglund, J., and Lundberg, A. (1991). Lekking in black grouse—A test of male viability. *Nature (London), 352*, 155–156.

Alatalo, R. V., Höglund, J., Lundberg, A., and Sutherland, W. J. (1992). Evolution of black grouse leks—Female preferences benefit males in larger leks. *Behav. Ecol., 3*, 53–59.

Alatalo, R. V., Lundberg, A., and Glynn, C. (1986b). Female pied flycatchers choose territory quality and not male characteristics. *Nature (London), 323*, 152–153.

Alcock, J. (1979a). The behavioural consequences of size variation among males of the territorial wasp *Hemipepsis ustulata* (Hymenoptera: Pompilidae). *Behaviour, 71*, 322–335.

Alcock, J. (1979b). The evolution of intraspecific diversity in male reproductive strategies in some bees and wasps. In M. S. Blum and N. A. Blum eds., *Sexual Selection and Reproductive Competition in Insects*, pp. 381–402. New York: Academic Press.

Alcock, J. (1981). Lek territoriality in a tarantula hawk wasp *Hempepsis ustulata* (Hymenoptera: Pompilidae). *Behav. Ecol. Sociobiol., 8*, 309–317.

Alcock, J. (1983). Territoriality by hilltopping males of the great purple hairstreak, *Atlides halesus* (Lepidoptera: Lyaenidae): Convergent evolution with a pompilid wasp. *Behav. Ecol. Sociobiol., 31*, 57–62.

Alcock, J., and Pyle, D. W. (1979). The complex courtship behavior of *Physiphora demandata* (F.) (Diptera: Otitidae). *Z. Tierpsychol., 49*, 352–362.

Alcock, J., and Smith, A. P. (1987). Hilltopping, leks and female choice in the carpenter bee *Xylocopa (Neoxylocopa) varipuncta. J. Zool., 211*, 1–10.

Alexander, R. D. (1975). Natural selection and specialized chorusing behaviour in acoustical insects. In D. Pimentel, ed., *Insects, Science and Society*. New York: Academic Press.

Aluja, M., Hendrichs, J., and Cabrera, M. (1983). Behavior and interactions between *Anastrepa ludens* (L.) and *A. Oblica* (M.) on a field caged mango tree. I. Lekking behavior and male territoriality. In R. Cavalloro, ed., *Fruit Flies of Economic Importance*, pp. 122–133. Rotterdam: A. A. Balkema.

Andersen, F. S. (1948). Contribution to the biology of the ruff (*Philomachus pugnax*), II. *Dansk Orn. For. Tidsskr., 42*, 125–148.

Andersson, M. (1982a). Female choice selects for extreme tail length in a widowbird. *Nature (London), 299*, 818–820.

Andersson, M. (1982b). Sexual selection, natural selection and quality advertisement. *Biol. J. Linn. Soc., 17*, 375–393.

Andersson, M. (1994). *Sexual Selection*. Princeton: Princeton University Press.

Andersson, M. B. (1986). Evolution of condition-dependent sex ornaments and mating preferences: Sexual selection based on viability differences. *Evolution, 40*, 804–816.

Andersson, M., and Norberg, Å. (1981). Evolution of reversed sexual size dimorphism and role partitioning among predatory birds, with a size scaling flight performance. *Biol. J. Linn. Soc., 15*, 105–130.

Andersson, S. (1989). Sexual selection and cues for female choice in leks of Jackson's widowbird *Euplectes jacksoni*. *Behav. Ecol. Sociobiol., 25*, 403–410.

Andersson, S. (1992). Female preferences for long tails in lekking Jackson's widowbirds: Experimental evidence. *Anim. Behav., 43*, 379–388.

Appleby, M. C. (1983). The probability of linearity in hierarchies. *Anim. Behav., 31*, 600–608.

Appolonio, M. (1989). Lekking in fallow deer: Just a matter of density? *Ethol. Ecol. Evol., 1*, 291–294.

Appolonio, M., Festa-Bianchet, M., and Mari, F. (1989a). Correlates of copulatory success in a fallow deer lek. *Behav. Ecol. Sociobiol., 25*, 89–97.

Appolonio, M., Festa-Bianchet, M., and Mari, F. (1989b). Effects of removal of successful males in a fallow deer lek. *Ethology, 83*, 320–325.

Appolonio, M., Festa-Bianchet, M., Mari, F., and Riva, M. (1990). Site-specific asymmetries in male copulatory success on a fallow deer lek. *Anim. Behav., 39*, 205–212.

Arak, A. (1982). Sneaky breeders. In C. J. Barnard, ed., *Producers and Scroungers: Strategies of Exploitation and Parasitism*, pp. 154–194. Beckenham, U.K.: Croom Helm.

Arak, A. (1983a). Male-male competition and mate choice in anuran amphibians. In P. Bateson, ed., *Mate Choice*, pp. 181–210. Cambridge, U.K.: Cambridge University Press.

Arak, A. (1983b). Sexual selection by male-male competition in natterjack toad choruses. *Nature (London), 306*, 261–262.

Arak, A. (1988a). Female mate selection in the natterjack toad: active choice or passive attraction? *Behav. Ecol. Sociobiol., 22*, 317–327.

Arak, A. (1988b). Sexual dimorphism in body size: A model and a test. *Evolution, 42*, 820–825.

Arak, A., and Eiriksson, T. (1992). Choice of singing sites by male bushcrickets (*Tettigonia viridissima*) in relation to signal propagation. *Behav. Ecol. Sociobiol., 30*, 365–372.

Arak, A., Eiriksson, T., and Radesäter, T. (1990). The adaptive significance of acoustic spacing in male bushcrickets *Tettigonia viridissima*: A peturbation experiment. *Behav. Ecol. Sociobiol., 26*, 1–7.

Arita, L. H., and Kaneshiro, K. Y. (1985). The dynamics of the lek system and mating success in males of the Mediterranean fruit fly, *Ceratitis capitata* (Weidemann). *Proc. Hawaii Entomol. Soc., 25*, 39–48.

Arnold, S. J. (1983). Sexual selection: The interface of theory and empiricism. In P. Bateson, ed., *Mate Choice*, pp. 67–107. Cambridge, U.K.: Cambridge University Press.

Avery, M. I. (1984). Lekking in birds: Choice, competition and reproductive constraints. *Ibis, 126*, 177–187.

Bailey, W. J., and Thiele, D. R. (1983). Male spacing behavior in the Tettigonidae: An experimental approach. In D. T. Gwynne and G. K. Morris, eds., *Orthopteran Mating Systems: Sexual Competition in a Diverse Group of Insects*, pp. 163–184. Boulder: Westview Press.

Baker, R. R. (1983). Insect territoriality. *Ann. Rev. Entomol., 28*, 65–89.

Ballard, W. B., and Robel, R. J. (1974). Reproductive importance of dominant male greater prairie chickens. *Auk, 91*, 75–85.

Balmford, A. (1990). Lekking in Uganda kob. Ph.D. thesis, Cambridge University.

Balmford, A., Albon, S., and Blakeman, S. (1992). Correlates of male mating success and female choice in a lek-breeding antelope. *Behav. Ecol., 3*, 112–123.

Balmford, A., Bartos, L., Brotherton, P., Herrmann, H., Lancingerova, J., Mika, J., and Zeeb, U. (1993). When to stop lekking: Density-related variation in the rutting behaviour of sika deer. *J. Zool. (London), 231*, 652–656.

Balmford, A., Deutsch, J. C., Nefdt, R.J.C., and Clutton-Brock, T. (1993). Testing hotspot models of lek evolution: Data from three species of ungulates. *Behav. Ecol. Sociobiol., 33*, 57–65.

Balmford, A., and Read, A. F. (1991). Testing alternative models of sexual selection through female choice. *TREE, 6*, 274–276.

Balmford, A., and Turyaho, M. (1992). Predation risk and lek-breeding in Uganda kob. *Anim. Behav., 44*, 117–127.

Barash, D. P. (1972). Lek behavior in the broad-tailed hummingbird. *Wilson Bull., 84*, 202–203.

Barnard, C. J., and Sibly, R. M. (1981). Producers and scroungers: A general model and its application to captive flocks of house sparrows. *Anim. Behav., 29*, 543–550.

Barnard, P., and Markus, M. B. (1989). Male copulation frequency and female competition for fertilizations in a promiscuous brood parasite, the pin-tailed whydah *Vidua macroura*. *Ibis, 131*, 421–425.

Basolo, A. (1990). Female preference for male sword length in the green swordtail *Xiphophorus helleri*. *Anim. Behav., 40*, 332–338.

Batra, S.W.T. (1979). Reproductive behavior of *Euaresta bella* and *E. festiva* (Diptera Tephritidae), potential agents for the biological control of adventive North American ragweeds (Ambrosia spp.) in Eurasia. *J. N.Y. Entomol. Soc., 87*, 118–125.

Baylis, J. R. (1981). The evolution of parental care in fishes, with reference to Darwin's rule of male sexual selection. *Env. Biol. Fish., 6*, 223–251.

Beani, L., and Turilazzi, S. (1990). Overlap at landmarks by lek-territorial and swarming males of two sympatric polistine wasps (Hymenoptera, Vespidae). *Ethol. Ecol. Evol., 2*, 419–431.

Beehler, B. M. (1987). Birds of paradise and mating system theory—Predictions and observations. *Emu, 87*, 78–89.

Beehler, B. M., and Foster, M. S. (1988). Hotshots, hotspots and female preferences in the organization of lek mating systems. *Am. Nat., 131*, 203–219.

Beehler, B., and Pruett-Jones, S. G. (1983). Display dispersion and diet of birds of paradise: A comparison of nine species. *Behav. Ecol. Sociobiol., 13*, 229–238.

Bell, G. (1978). The handicap principle in sexual selection. *Evolution, 32*, 872–885.

Bendell, J. F., and Elliot, P. W. (1967). *Behaviour and the Regulation of Numbers in Blue Grouse*. Canadian Wildlife Service Reprint Series.

Berger, D. D., Hamerstrom, F., and Hamerstrom, F. N., Jr. (1963). The effect of raptors on prairie chickens on booming grounds. *J. Wildl. Manag., 27*, 778–791.

Bertram, B.C.R. (1978). Living in groups: Predators and prey. In J. R. Krebs and N. B. Davies, eds., *Behavioural Ecology: An Evolutionary Approach*, pp. 64–96. Oxford: Blackwell.

Bertram, B.C.R. (1980). Vigilance and group size in ostriches. *Anim. Behav., 28*, 278–286.

Beuchner, H. K. (1961). Territorial behavior in Uganda kob. *Science, 133*, 698–699.

Bikchandani, S., Hirschleifer, D., and Welch, I. (1992). A theory of fads, fashion, custom and cultural change as information cascades. *J. Polit. Econ., 100*, 992–1026.

Birkhead, T. R. (1988). Behavioral aspects of sperm competition in birds. *Adv. Study Behav., 18*, 35–72.

Birkhead, T. R., Atkin, L., and Møller, A. P. (1987). Copulation behaviour in birds. *Behaviour, 101*, 101–138.

Birkhead, T. R., and Møller, A. P. (1992). *Sperm Competition in Birds: Evolutionary Causes and Consequences*. London: Academic Press.

Birkhead, T. R., and Møller, A. P. (1993). Female control of paternity. *TREE, 8*, 100–104.

Bleiweiss, R. (1985). Iridescent polychromatism in a female hummingbird: Is it related to feeding strategies? *Auk, 102*, 701–713.

Boag, D. A., and Sumanik, K. M. (1969). Characteristics of drumming sites selected by ruffed grouse in Alberta. *J. Wildl. Manag., 33*, 621–628.

Borgia, G. (1979). Sexual selection and the evolution of mating systems. In M. S. Blum and N. A. Blum eds., *Sexual Selection and Reproductive Competition in Insects*, pp. 19–80. New York: Academic Press.

Borgia, G., Pruett-Jones, S. G., and M. A. Pruett-Jones, (1985). The evolution of bower building and the assessment of male quality. *Z. Tierpsychol., 76*, 225–236.

Bourne, G. R. (1992). Lekking behavior in the neotropical frog *Ololygon rubra*. *Behav. Ecol. Sociobiol., 31*, 173–180.

Bowen, S. H. (1984). Differential habitat utilisation of *Sarotherodon mossambicus* in Lake Valencia: Significance for fitness. *J. Fish. Biol., 24*, 115–121.

Bradbury, J. W. (1977). Lek mating behavior in the hammerheaded bat. *Z. Tierpsychol., 45*, 225–255.

Bradbury, J. W. (1981). The evolution of leks. In R. D. Alexander and D. W. Tinkle, eds., *Natural Selection and Social Behaviour*, pp. 138–169. New York and Concord: Chiron Press.

Bradbury, J. W. (1985). Contrasts between insects and vertebrates in the evolution of male display, female choice and lek mating. *Fort. Zool., 31*, 273–289.

Bradbury, J. W., and Andersson, M. B., eds. (1987). *Sexual Selection: Testing the Alternatives*. Chichester, U.K.: John Wiley.

Bradbury, J. W., and Gibson, R. M. (1983). Leks and mate choice. In P. Bateson, ed., *Mate Choice*, pp. 109–138. Cambridge, U.K.: Cambridge University Press.

Bradbury, J. W., Gibson, R. M., McCarthy, C. E., and Vehrencamp, S. L. (1989a). Dispersion of displaying male sage grouse. II. The role of female dispersion. *Behav. Ecol. Sociobiol., 24*, 15–24.

Bradbury, J. W., Gibson, R. M., and Tsai, I. M. (1986). Hotspots and the dispersion of leks. *Anim. Behav., 34*, 1694–1709.

Bradbury, J. W., and Vehrencamp, S. L. (1977). Social organization and foraging in Emballonurid bats. III. Mating systems. *Behav. Ecol. Sociobiol., 2*, 1–17.

Bradbury, J. W., Vehrencamp, S. L., and Gibson, R. (1985). Leks and the unanimity of female choice. In P. J. Greenwood, P. H. Harvey, and M. Slatkin, eds., *Evolution: Essays in Honour of John Maynard Smith*, pp. 301–314. Cambridge, U.K.: Cambridge University Press.

Bradbury, J. W., Vehrencamp, S. L., and Gibson, R. M. (1989b). Dispersion of displaying male sage grouse. I. Patterns of temporal variation. *Behav. Ecol. Sociobiol., 24*, 1–14.

Brittain, J. E. (1982). Biology of mayflies. *Ann. Rev. Entomol., 27*, 119–147.

Brockman, H. J. (1990). Mating behaviour of horseshoe crabs, *Limulus polyphemus. Behaviour, 114*, 206–220.

Brooks, D. R., and McLennan, D. A. (1991). *Phylogeny, Ecology, and Behavior: A Research Program*. Chicago: University of Chicago Press.

Brosset, A. (1982). The social life of the African forest yellow-whiskered greenbul *Andropadus latirostris. Z. Tierpsychol., 60*, 239–255.

Brown, J. L. (1987). *Helping and Communal Breeding in Birds*. Princeton: Princeton University Press.

Brüll, H. (1961). Birkwildforschung und Birkwildhege in Schleswig-Holstein. *Z. Jagdwiss., 7*, 104–126.

Bucher, T. L., Ryan, M. J., and Bartholomew, G. A. (1982). Oxygen consumption during resting, calling and nest building in the frog, *Physalaemus pustulosus. Physiol. Zool., 55*, 10–22.

Buechner, H. K. (1961). Territorial behavior in Uganda kob. *Science, 133*, 698–699.

Buechner, H. K., and Schloeth, R. (1965). Ceremonial mating behaviour in Uganda kob *(Adenota kob thomasi* Neuman). *Z. Tierpsychol., 22*, 209–225.

Burk, T. (1982). Evolutionary significance of predation on sexually signalling males. *Fla. Entomol., 65*, 90–105.

Burla, H. (1990). Lek behaviour in hypercephalic *Zygothrica dispar* Wiedemann (Diptera, Drosophiladae). *Z. Zool. Syst. Evol. Forsch., 28*, 69–77.

Burley, N. (1986). Sexual selection for aesthetic traits in species with parental care. *Am. Nat., 127*, 415–445.

Burley, N. (1988). The differential-allocation hypothesis: An experimental test. *Am. Nat., 132*, 611–628.

Butlin, R. K., Woodhatch, C. W., and Hewitt, G. M. (1987). Male spermatophore investment increases female fecundity in a grasshopper. *Evolution, 41*, 221–225.

Cade, T. J. (1960). Ecology of the peregrine and gyrfalcon populations in Alaska. *Univ. Calif. Publ. Zool., 63*, 151–290.

Cade, W. (1975). Acoustically orienting parasitoids: Fly phonotaxis to cricket song. *Science, 190*, 1312–1313.

Cade, W. (1979). The evolution of alternative reproductive strategies in field crickets. In M. S. Blum and N. A. Blum, eds., *Sexual Selection and Reproductive Competition in Insects*, pp. 343–379. New York: Academic Press.

Cade, W. H. (1981). Field cricket spacing, and the phonotaxis of crickets and parasitoid flies to clumped and isolated cricket song. *Z. Tierpsychol., 55*, 365–375.

Cade, W., and Wyatt, D. R. (1984). Factors affecting calling behaviour in field crickets, *Teleogryllus* and *Gryllus* (age, weight, density and parasites). *Behaviour, 88*, 61–75.

Campanella, P. J., and Wolf, L. L. (1973). Temporal leks as a mating system in a temperate zone dragonfly (Odonata: Anisoptera). I. *Plathemis lydia* (Drury). *Behaviour, 51*, 49–87.

Cartar, R. V., and Lyon, B. E. (1988). The mating system of the buff-breasted sandpiper: Lekking and resource defense polygyny. *Ornis Scand., 19*, 74–76.

Chapman, D. I., and Chapman, N. (1975). *Fallow Deer: Their History, Distribution and Biology*. Lavenham, U.K.: Terence Dalton.

Charlesworth, B. (1987). The heritability of fitness. In J. W. Bradbury and M. B. Andersson eds., *Sexual Selection: Testing the Alternatives*, pp. 21–40. Chichester, U.K.: John Wiley.

Cherry, M. I. (1992). Sexual selection in the leopard toad, *Bufo pardalis. Behaviour, 120*, 164–176.

Cherry, M. I. (1993). Sexual selection in the raucous toad, *Bufo rangeri. Anim. Behav., 45*, 359–373.

Christy, J. H. (1978). Adaptive significance of reproductive cycles in the fiddler crab *Uca pugilator*: A hypothesis. *Science, 199*, 453–455.

Christy, J. H. (1982). Burrow structure and use in the sand fiddler crab, *Uca pugilator* (Bosc). *Anim. Behav., 30*, 687–694.

Clavijo, I. E. (1983). Pair spawning and formation of a lek-like mating system in the parrotfish *Scarus vetula. Copeia, 1983*, 253–256.

Clutton-Brock, T. H. (1989). Mammalian mating systems. *Proc. Roy. Soc. London B, 236*, 339–372.

Clutton-Brock, T. H., Deutsch, J. C., and Nefdt, R.J.C. (1993). The evolution of ungulate leks. *Anim. Behav., 46*, 1121–1138.

Clutton-Brock, T. H., Green, D., Hiraiwa-Hasegawa, M., and Albon, S. D. (1988). Passing the buck: Resource defence, lek breeding and mate choice in fallow deer. *Behav. Ecol. Sociobiol., 23*, 281–296.

Clutton-Brock, T. H., Guinness, F. E., and Albon, S. D. (1982). *Red Deer: Behavior and Ecology of Two Sexes*. Edinburgh, U.K.: Edinburgh University Press.

Clutton-Brock, T. H., Harvey, P. H., and Rudder, P. (1977). Sexual dimorphism, socioeconomic sex ratio, and body weight in primates. *Nature (London)*, 797–800.

Clutton-Brock, T. H., Hiraiwa-Hasegawa, M., and Robertson, A. (1989). Mate choice on fallow deer leks. *Nature (London), 340*, 463–465.

Clutton-Brock, T., and McComb, K. (1993). Experimental tests of copying and mate choice in fallow deer (*Dama dama*). *Behav. Ecol., 4*, 191–194.

Clutton-Brock, T. H., Price, O., and MacColl, A. (1992). Mate retention, harassment and the evolution of ungulate leks. *Behav. Ecol., 3*, 234–242.

Coe, M. (1966). The biology of *Tilapia grahami* Boulenger in Lake Magadi, Kenya. *Acta Trop., 23*, 146–177.

Constantz, G. D. (1975). Behavioral ecology of mating in the male gila topminnow, *Poeciliopsis occidentalis* (Cyprinodontiformes: Poeciliidae). *Ecology, 56*, 966–973.

Cooper, W. T., and Forshaw, J. M. (1979). *The Birds of Paradise and Bowerbirds*. Sydney: Collins.

Corbet, G. B., and Hill, J. E. (1991). *A World List of Mammalian Species*. 3d ed. Oxford: Oxford University Press.

Courtney, S. P., and Parker, G. A. (1985). Mating behaviour of the tiger blue butterfly (*Tarucus theophrastus*): Competitive mate-searching when not all females are captured. *Behav. Ecol. Sociobiol., 17*, 213–221.

Cox, C. R., and LeBouef, B. J. (1977). Female incitation of male competition: A mechanism in sexual selection. *Am. Nat., 111*, 317–335.

Cramp, S., ed. (1983). *Handbook of the Birds of Europe, the Middle East and North Africa*. Vol. 3. Oxford: Oxford University Press.

Cramp, S., and Simmons, K.E.L., eds. (1980). The birds of the western palearctic. Vol. 2. Oxford: Oxford University Press.

Crook, J. H. (1964). The evolution of social organization and visual communication in the weaver birds (Ploceinae). *Behaviour Suppl., 10*, 1–178.

Crook, J. H. (1972). Sexual selection, dimorphism, and social organization in the primates. In B. Campbell, ed., *Sexual Selection and the Descent of Man*, pp. 1871–1971. Chicago: Aldine.

Darwin, C. (1871). *The Descent of Man and Selection in Relation to Sex*. London: John Murray.

Davies, N. B. (1978). Ecological questions about territorial behaviour. In J. R. Krebs and N. B. Davies, eds., *Behavioural Ecology: An Evolutionary Approach*, pp. 317–350. Oxford: Blackwell.

Davies, N. B. (1985). Cooperation and conflict among dunnocks, *Prunella modularis*, in a variable mating system. *Anim. Behav., 33*, 628–648.

Davies, N. B. (1986). Reproductive success of dunnocks *Prunella modularis* in a variable mating system. I. Factors influencing provisioning rate, nestling weight and fledging success. *J. Anim. Ecol., 55*, 123–138.

Davies, N. B. (1991). Mating systems. In J. R. Krebs and N. B. Davies, *Behavioural Ecology*. 3d ed. Oxford: Blackwell.

Davies, N. B., and Houston, A. I. (1986). Reproductive success of dunnocks *Prunella modularis* in a variable mating system. II. Conflicts of interest among breeding adults. *J. Anim. Ecol., 55*, 139–154.

Davis, T.A.W. (1949). Display of white-throated manakins *Corapipo gutturalis*. *Ibis, 91*, 146–147.

Davis, T.A.W. (1958). The displays and nests of three forest hummingbirds in British Guyana. *Ibis, 100*, 31–39.

Davison, G.W.H. (1981). Sexual selection and the mating system of *Argusianus argus* (Aves: Phasianidae). *Biol. J. Linn. Soc., 15*, 91–104.

Dawkins, R. (1980). Good strategy or evolutionarily stable strategy? In G. W. Barlow and S. Silverberg, eds., *Sociobiology: Beyond Nature/Nurture?*, pp. 331–367. Boulder: Westview Press.

de Lope, F., and Møller, A. P. (1993). Female reproductive effort depends on the degree of ornamentation of their mates. *Evolution, 47*, 1152–1160.

de Schauensee, M. R., Phelps, W. H., Jr., and Tudor, G. (1978). *Birds of Venezuela*. Princeton: Princeton University Press.

Deutsch, J. C. (1992) Reproductive strategies in a lek-breeding antelope, the Uganda kob. Ph.D. thesis, Cambridge University.

Deutsch, J. C. (1994a). Lekking by default: Female habitat preferences and male strategies in Uganda kob. *J. Anim. Ecol., 63*, 101–115.

Deutsch, J. C. (1994b.) Uganda kob mating success does not increase on larger leks. *Behav. Ecol. Sociobiol., 34*, 451–459.

Deutsch, J. C., and Nefdt, R.J.C. (1992). Olfactory cues influence female choice in two lek-breeding antelopes. *Nature (London), 356*, 596–598.

Deutsch, J. C., and Weeks, P. (1992). Uganda kob prefer high visibility leks and territories. *Behav. Ecol., 3*, 223–233.

de Vos, G. J. (1979). Adaptedness of arena behaviour in black grouse (*Tetrao tetrix*) and other grouse species (Tetraonidae). *Behaviour, 68*, 277–314.

de Vos, G. J. (1983). Social behaviour of black grouse: An observational and experimental field study. *Ardea, 71*, 1–103.

DeVries, P. J. (1978). Observations on the apparent lek behaviour in Costa Rican rainforest *Perrhybris pyrrha* Cramer (Pieridae). *J. Res. Lepid., 17*, 142–144.

Dodson, G. (1982). Mating and territoriality in wild *Anastrepha suspensa* (Diptera: Tephritidae) in field cages. *J. Ga. Entomol. Soc., 17*, 189–200.

Dodson, G. (1986). Lek mating system and large male aggressive advantage in a gall-forming tephritid fly (Diptera: Tephritidae). *Ethology, 72*, 99–108.

Doolan, J. M., and MacNally, R. C. (1981). Spatial dynamics and breeding ecology in the cicada *Cytosoma saundersii*: The interaction between distributions of resources and intraspecific behaviour. *J. Anim. Ecol., 50*, 925–940.

Downes, J. A. (1969). The swarming and mating flight of Diptera. *Ann. Rev. Entomol., 14*, 271–298.

Drent, R. H., and Daan, S. (1980). The prudent parent: Energetic adjustments in avian breeding. *Ardea, 68*, 225–252.

Droney, D. C. (1992). Sexual selection in a lekking Hawaiian *Drosophila*: The roles of male competition and female choice in male mating success. *Anim. Behav., 44*, 1007–1020.

Droney, D. C. (1994). Tests of hypotheses for lek formation in a Hawaiian *Drosophila*. *Anim. Behav., 47*, 351–361.

Dunning, J. B. (1992). *CRC Handbook of Avian Body Mass*. Boca Raton, Fla.: CRC Press.

Eldredge, N., and Cracraft, J. (1980). *Phylogenetic Patterns and the Evolutionary Process*. New York: Columbia University Press.

Emlen, S. T. (1968). Territoriality in the bullfrog *Rana catesbeiana*. *Copeia, 1968*, 240–243.

Emlen, S. T. (1976). Lek organization and mating strategies in the bullfrog. *Behav. Ecol. Sociobiol., 1*, 283–313.

Emlen, S. T., and Oring, L. W. (1977). Ecology, sexual selection and the evolution of mating systems. *Science, 197*, 215–223.

Endler, J. A. (1987). Predation, light intensity, and courtship behaviour in *Poecilia reticulata*. *Anim. Behav., 35*, 1376–1385.

Endler, J. A. (1991). Variation in the appearance of guppy color patterns to guppies and their predators under different light conditions. *Vision Res., 31*, 587–608.

Endler, J. A. (1992). Signals, signal conditions, and the direction of evolution. *Am. Nat., 139*, S125–S153.

Enquist, M. (1985). Communication during aggressive interactions with particular reference to variation in choice of behaviour. *Anim. Behav., 33*, 1152–1161.

Enquist, M., and Arak, A. (1993). Selection of exaggerated male traits by female aesthetic senses. *Nature (London), 361*, 446–448.

Eriksson, D. (1991) The significance of song for species recognition and mate choice in the pied flycatcher *Ficedula hypoleuca*. Ph.D. thesis, Uppsala University.

Espmark, Y., and Brunner, W. (1974). Observations on rutting behaviour in fallow deer (*Dama dama*). *Säugetierkundl. Mitt., 22*, 135–142.

Estes, R. D. (1968). Territorial behaviour of the wildebeest (*Connochaetes taurinus* Burchell, 1823). *Z. Tierpsychol., 26*, 284–370.

Falconer, D. S. (1981). *Introduction to Quantitative Genetics*. 2d ed. London: Longman.

Fay, F. H., Ray, G. C., and Kibalchich, A. A. (1984). Time and location of mating and associated behaviour of the Pacific walrus *Odobenus rosmarus divergens* Illiger. In F. H. Fay and G. A. Fedoseev, eds., *Soviet-American Cooperative Research on Marine Mammals*, pp. 89–99. NOAA Technical Report, NMFS 12.

Feller, W. (1968). *An Introduction to Probability Theory and Its Applications*. 3d ed. New York: John Wiley.

Fellers, G. M. (1979). Mate selection in the gray treefrog, *Hyla versicolor*. *Copeia, 1979*, 286–290.

Felsenstein, J. (1985). Phylogenies and the comparative method. *Am. Nat., 125*, 1–15.

Fernald, R. D., and Hirata, N. R. (1977). Field study of Haplochromis burtoni: Quantitative behavioural observations. *Anim. Behav., 25*, 643–653.

Fisher, R. A. (1930). *The Genetical Theory of Natural Selection*. 1st ed. Oxford: Clarendon Press.

Fisher, R. A. (1958). *The Genetical Theory of Natural Selection*. 2d ed. New York: Dover.

Fiske, P. (1994). Sexual selection in the lekking great snipe (*Gallinago media*): Male mating success and female behaviour at the lek. Ph.D. thesis, University of Trondheim.

Fiske, P., and Kålås, J. A. (In prep.). Mate sampling and copulation behaviour of great snipe females. *Anim. Behav.*

Fleishman, L. (1992). The influence of the sensory system and the environment on motion patterns in the visual displays of anoline lizards and other vertebrates. *Am. Nat., 139*, S36–S61.

Floody, O. R., and Arnold, A. P. (1975). Uganda kob (*Adenota kob thomasi*): Territoriality and the spatial distribution of sexual and antagonistic behaviour at a territorial ground. *Z. Tierpsychol., 37*, 192–212.

Forsgren, E. (1993). Predation risk affects mate choice in a gobiid fish. *Am. Nat., 140*, 1041–1049.

Foster, M. S. (1977). Odd couples in manakins: A study of social organization and cooperative breeding in *Chiroxiphia linearis*. *Am. Nat., 111*, 845–853.

Foster, M. S. (1981). Cooperative behaviour and social organization of the swallow-tailed manakin (*Chiroxiphia caudata*). *Behav. Ecol. Sociobiol., 9*, 167–177.

Foster, M. S. (1983). Disruption, dispersion, and dominance in lek-breeding birds. *Am. Nat., 122*, 53–72.

Fretwell, S. D. (1972). *Populations in a Seasonal Environment*. Princeton: Princeton University Press.

Fretwell, S. D., and Lucas, H. L., Jr. (1970). On territorial behaviour and other factors influencing habitat distribution in birds. *Acta Biotheor., 19*, 16–36.

Fryer, G., and Iles, T. D. (1972). *The Cichlid Fishes of the Great Lakes of Africa*. Edinburgh, U.K.: Oliver and Boyd.

Fryxell, J. M. (1987). Lek breeding and territorial aggression in white-eared kob. *Ethology, 75*, 211–220.

Gadamer, H. (1857). Das Balzen der *Scolopax major*. *J. Ornithol., 6*, 235–236.

Geist, V. (1974). On the relationship of social evolution and ecology in ungulates. *Am. Zool., 14*, 205–220.

Gewalt, W. (1959). *Die Grosstrappe (Otis tarda L.)*. Wittenberg, Germany: A. Ziemsen.

Gibbs, H. L., Weatherhead, P. J., Boag, P. T., White, B. N., Tabak, L. M., and Hoysak, D. J. (1990). Realized reproductive success of polygynous red-winged blackbirds revealed by DNA markers. *Science, 250*, 1394–1397.

Gibson, R. M. (1989). Field playback of male display attracts females in lek-breeding sage grouse. *Behav. Ecol. Sociobiol., 24*, 439–443.

Gibson, R. M. (1990). Relationships between blood parasites, mating success and phenotypic cues in male sage grouse *Centrocercus urophasianus*. *Am. Zool., 30*, 271–276.

Gibson, R. M. (1992). Lek formation in sage grouse: The effect of female choice on male territory settlement. *Anim. Behav., 43*, 443–450.

Gibson, R. M., and Bachman, G. C. (1992). The costs of female choice in a lekking bird. *Behav. Ecol., 3*, 300–309.

Gibson, R. M., and Bradbury, J. W. (1985). Sexual selection in lekking sage grouse: Phenotypic correlates of male mating success. *Behav. Ecol. Sociobiol., 18*, 117–123.

Gibson, R. M., and Bradbury, J. W. (1986). Male and female mating strategies on sage grouse leks. In D. I. Rubenstein and R. W. Wrangham, eds., *Ecological Aspects of Social Evolution*, pp. 379–398. Princeton: Princeton University Press.

Gibson, R. M., and Bradbury, J. W. (1987). Lek organization in sage grouse: Variations on a territorial theme. *Auk, 104*, 77–84.

Gibson, R. M., Bradbury, J. W., and Vehrencamp, S. L. (1991). Mate choice in lekking sage grouse revisited: The roles of vocal display, female site fidelity, and copying. *Behav. Ecol., 2*, 165–180.

Gibson, R. M., and Guinness, F. E. (1980). Behavioural factors affecting male reproductive success in red deer (*Cervus elaphus*). *Anim. Behav., 28*, 1163–1174.

Gibson, R. M., and Höglund, J. (1992). Copying and sexual selection. *TREE, 7*, 229–232.

Gibson, R. M., Taylor, C. E., and Jefferson, D. R. (1990). Lek formation by female choice: A simulation study. *Behav. Ecol., 1*, 36–42.

Gilliard, E. T. (1969). *Birds of Paradise*. New York: Natural History Press.

Göransson, G., von Schantz, T., Fröberg, I., Helgée, A., and Wittzell, H. (1990). Male characteristics, viability and harem size in the pheasant. *Anim. Behav., 40*, 89–104.

Gosling, L. M. (1986). The evolution of mating strategies in male antelopes. In D. I. Rubenstein and R. W. Wrangham, eds., *Ecological Aspects of Social Evolution*. Princeton: Princeton University Press.

Gosling, L. M., and Petrie, M. (1990). Lekking in topi: A consequence of satellite behaviour by small males at hotspots. *Anim. Behav., 40*, 272–287.

Gosling, L. M., Petrie, M., and Rainey, M. E. (1987). Lekking in topi: A high cost, specialist strategy. *Anim. Behav., 35*, 616–618.

Götmark, F., and Andersson, M. (1984). Colonial breeding reduces nest predation in the common gull (*Larus canus*). *Anim. Behav., 32*, 485–492.

Gottlander, K. (1987). Variation in the song rate of the male pied flycatcher *Ficedula hypoleuca*: Causes and consequences. *Anim. Behav., 35*, 1037–1043.

Grafen, A. (1989). The phylogenetic regression. *Trans. Roy. Soc. London B, 326*, 119–157.

Grafen, A. (1990a). Biological signals as handicaps. *J. theor. Biol.*, 517–546.

Grafen, A. (1990b). Sexual selection unhandicapped by the Fisher process. *J. theor. Biol., 144*, 473–516.

Grant, B. R. (1990). The significance of subadult plumage in Darwin's finches, *Geospiza fortis*. *Behav. Ecol., 1*, 161–170.

Gratson, M. W., Gratson, G. K., and Bergerud, A. T. (1991). Male dominance and copulation disruption do not explain variance in male mating success on sharp-tailed grouse (*Tympanuchus phasianellus*) leks. *Behaviour, 118*, 187–213.

Greenfield, M. D., and Shaw, K. C. (1983). Adaptive significance of chorusing with special reference to the Orthoptera. In D. T. Gwynne and G. K. Morris, eds., *Orthopteran Mating Systems: Sexual Competition in a Diverse Group of Insects*, pp. 1–27. Boulder: Westview Press.

Greenfield, M. D., Shelly, T. E., and Downum, K. R. 1987. Variation in host plant quality: Implications for territoriality in a desert grasshopper. *Ecology, 68*, 828–838.

Greenspan, B. N. (1980). Male size and reproductive success in the communal courtship system of the fiddler crab *Uca rapax*. *Anim. Behav., 28*, 387–392.

Gross, M. R. (1991). Evolution of alternative reproductive strategies: Frequency dependent sexual selection in male bluegill sunfish. *Phil. Trans. Roy. Soc. London B, 332*, 59–66.

Gustafsson, L. (1986). Lifetime reproductive success and heritability: Empirical support for Fisher's fundamental theorem. *Am. Nat.*, *128*, 761–764.

Gustafsson, L., and Pärt, T. (1990). Acceleration of senescence in the collared flycatcher *Ficedula albicollis* by reproductive costs. *Nature (London)*, *347*, 279–281.

Gwynne, D. T. (1984). Courtship feeding increases female reproductive success in bushcrickets. *Nature (London)*, *307*, 361–363.

Gwynne, D. T., and Morris, G. K., eds. (1983). *Orthopteran Mating Systems: Sexual Competition in a Diverse Group of Insects.* Boulder: Westview Press.

Halliday, T. (1983). Do frogs and toads choose their mates? *Nature (London)*, *306*, 226–227.

Hamerstrom, F. N., Jr., and Hamerstrom, F. (1955). Population density and behavior in Wisconsin prairie chickens *(Tympanuchus cupido pinnatus)*. In *Proc. 11th Int. Ornith. Congr.*, pp. 459–466.

Hamerstrom, F., and Hamerstrom, F. (1958). Comparability of some social displays of grouse. In *Proc. 12th Int. Ornith. Congr.*, pp. 274–293.

Hamerstrom, F., and Hamerstrom, F. (1973). The prairie chicken in Wisconsin. *Wis. Dept. Nat. Res., Tech. Bull., 64*, 1–52.

• Hamilton, W. D. (1970). Selfish and spiteful behaviour in an evolutionary model. *Nature (London)*, *228*, 1218–1220.

Hamilton, W. D. (1971). Geometry of the selfish herd. *J. theor. Biol., 7*, 1–52.

Hamilton, W. D., and Zuk, M. (1982). Heritable true fitness and bright birds: A role for parasites. *Science, 218*, 384–386.

Hammerstein, P., and Parker, G. A. (1987). Sexual selection: Games between the sexes. In J. W. Bradbury and M. B. Andersson, eds., *Sexual Selection: Testing the Alternatives*, pp. 119–142. Chichester, U.K.: John Wiley.

Hartzler, J. E. (1974). Predation and the daily timing of sage grouse leks. *Auk, 91*, 532–536.

Hartzler, J. E., and Jenni, D. A. (1988). Mate choice by female sage grouse. In A. T. Bergerud and M. W. Gratson, eds., *Adaptive Strategies and Population Ecology of Northern Grouse*, pp. 240–269. Minneapolis: University of Minnesota Press.

Harvey, P. H., and Bradbury, J. W. (1991). Sexual selection. In J. R. Krebs and N. B. Davies, eds., *Behavioural Ecology: An Evolutionary Approach*, pp. 203–233. Oxford: Blackwell.

Harvey, P. H., and Pagel, M. D. (1991). *The Comparative Method in Evolutionary Biology.* Oxford: Oxford University Press.

Hassal, J. R., and Zaveri, K. (1979). *Acoustic Noise Measurements.* Søborg: Bruel and Kjaer.

Hassel, M. P., and Varley, G. C. (1969). New inductive population model for insect parasites and its bearing on biological control. *Nature (London)*, *223*, 1133–1136.

Hasson, O. (1989). Amplifiers and the handicap principle in sexual selection: A different emphasis. *Proc. Roy. Soc. London B, 235*, 383–406.

Hasson, O. (1990). The role of amplifiers in sexual selection: An integration of the amplifying and the Fisherian mechanisms. *Evol. Ecol., 4*, 277–289.

Hasson, O. (1991). Sexual displays as amplifiers: Practical examples with an emphasis on feather decorations. *Behav. Ecol., 2*, 189–197.

Hastings, P. A. (1986). Female choice and male reproductive success in the angel blenny, *Coralliozetus angelica* (Teleostei: Chaenopsidae). *Anim. Behav., 36*, 115–124.

Hedlund, L., and Robertson, J.G.M. (1989). Lekking behaviour in crested newts, *Triturus cristatus. Ethology, 80*, 111–119.

Heisler, I. L. (1984). A quantitative genetic model for the origin of mating preferences. *Evolution, 38*, 1283–1295.

Heisler, I. L., and Curtsinger, J. W. (1990). Dynamics of sexual selection in diploid populations. *Evolution, 44*, 1164–1176.

Hendrichs, J., and Reyes, S. (1987). Reproductive behavior and post-mating female guarding in the monophagous multivoltine *Dacus longistylus* (Wied.) (Diptera: Tephritidae) in southern Egypt. In A. P. Economopoulos, ed., *Second International Symposium on Fruit Flies*, pp. 303–313. Colombari, Greece: Elsevier.

Hennig, W. 1966. *Phylogenetic Systematics*. Urbana: Univeristy of Illinois Press.

Hibino, Y. (1986). Female choice for male gregariousness in a stink bug, *Megacopta punctatissimum* (Montadon) (Heterpotera, Plataspidae). *J. Ethol., 4*, 91–95.

Hill, G. (1991). Plumage coloration is a sexually selected indicator of male quality. *Nature (London), 350*, 337–339.

Hill, W. L. (1991). Correlates of male mating success in the ruff *Philomachus pugnax*, a lekking shorebird. *Behav. Ecol. Sociobiol., 29*, 367–372.

Hjorth, I. (1970). Reproductive behaviour in Tetraonidae. *Swedish Wildlife (Viltrevy), 7*, 190–596.

Hodosh, R. J., Ringo, J. M., and McAndrew, F. T. (1979). Density and lek displays in *Drosophila grimshawi* (Diptera: Drosophilidae). *Z. Tierpsychol., 49*, 164–172.

Hoelzer, G. A. (1989). The good parent process of sexual selection. *Anim. Behav., 38*, 1067–1078.

Hoffman, A. A., and Blows, M. W. (1992). Evidence that *Drosophila mycetophaga* Malloch (Diptera: Drosophilidae) is not a true "lekking" species. *J. Aust. Entomol. Soc., 31*, 219–221.

Hogan-Warburg, A. J. (1966). Social behavior of the ruff, *Philomachus pugnax* (L.). *Ardea, 54*, 109–229.

Höglund, J. (1988). Pairing and spawning patterns in the common toad *Bufo bufo*: The effects of sex ratios and the time available for male-male competition. *Anim. Behav., 38*, 423–429.

Höglund, J. (1989). Size and plumage dimorphism in lek-breeding birds: A comparative analysis. *Am. Nat., 134*, 72–87.

Höglund, J. (1993). Costs of mate advertisement. *Etologia, 3*, 143–150.

Höglund, J., Alatalo, R. V., Gibson, R. M., and Lundberg, A. (In press). Mate choice copying in black grouse. *Anim. Behav.*

Höglund, J., Alatalo, R. V., and Lundberg, A. (1990b). Copying the mate choice of others? Observations on female black grouse. *Behaviour, 114*, 221–231.

Höglund, J., Alatalo, R. V., and Lundberg, A. (1992b). The effects of parasites on male ornaments and female choice in the lek-breeding black grouse (*Tetrao tetrix*). *Behav. Ecol. Sociobiol., 30*, 71–76.

Höglund, J., Alatalo, R. V., Lundberg, A., and Rätti, O. 1994. Context-dependent effects of tail-ornament damage on mating success in black grouse. *Behav. Ecol., 5*, 182–187.

Höglund, J., Eriksson, M., and Lindell, L.-E. (1990a). Females of the lek-breeding great snipe, *Gallinago media,* prefer males with white tails. *Anim. Behav., 40,* 23–32.

Höglund, J., Kålås, J.-A., and Løfaldli, L. (1990c). Sexual dimorphism in the lekking great snipe. *Ornis Scand., 21,* 1–6.

Höglund, J., Kålås, J.-A., and Fiske, P. (1992a). The costs of secondary sexual characters in the lekking great snipe (*Gallinago media*). *Behav. Ecol. Sociobiol., 30,* 309–315.

Höglund, J., and Lundberg, A. (1987). Sexual selection in a monomorphic lekbreeding bird: Correlates of male mating success in the great snipe *Gallinago media. Behav. Ecol. Sociobiol., 21,* 211–216.

Höglund, J., and Lundberg, A. (1989). Plumage color correlates with body size in the ruff (*Philomachus pugnax*). *Auk, 106,* 336–338.

Höglund, J., Montgomerie, R., and Widemo, F. (1993). Costs and consequences of variation in the size of ruff leks. *Behav. Ecol. Sociobiol., 32,* 31–39.

Höglund, J., and Robertson, J.G.M. (1988). Chorusing behaviour, a density-dependent alternative mating strategy in male common toads (*Bufo bufo*). *Ethology, 79,* 324–332.

Höglund, J., and Robertson, J.G.M. (1990a). Spacing of leks in relation to female home ranges, habitat requirements and male attractiveness in the great snipe (*Gallinago media*). *Behav. Ecol. Sociobiol., 26,* 173–180.

• Höglund, J., and Robertson, J.G.M. (1990b). Female preferences, male decision rules and the evolution of leks in the great snipe *Gallinago media. Anim. Behav., 40,* 15–22.

Höglund, J., and Säterberg, L. (1989). Sexual selection in common toads: Correlates with age and body size. *J. Evol. Biol., 2,* 367–372.

Höglund, J., and Sillén-Tullberg, B. (1994). Does lekking promote the evolution of male-biased size dimorphism in birds? On the use of comparative approaches. *Am. Nat. 144,* 881–889.

Horn, H. S. (1968). The adaptive significance of colonial nesting in Brewer's blackbird *Euphagus cyanocephalus. Ecology, 49,* 682–694.

Hosmer, D. W., Jr., and Lemeshow, S. (1989). *Applied Logistic Regression.* New York: John Wiley.

Howard, R., and Moore, A. (1991). *A Complete Checklist of the Birds of the World.* 2d ed. London: Academic Press.

Howard, R. D. (1978a). The evolution of mating strategies in bullfrogs, *Rana catesbeiana. Evolution, 32,* 850–871.

Howard, R. D. (1978b). The influence of male-defended oviposition sites on early embryo mortality in bullfrogs. *Ecology, 59,* 789–798.

Howard, R. D. (1983). Sexual selection and variation in reproductive success in a long-lived organism. *Am. Nat., 122,* 301–325.

Howard, R. D. (1984). Alternative mating behaviours of young male bullfrogs. *Amer. Zool. 24,* 397–406.

Huxley, J. S. (1938a). Darwin's theory of sexual selection and the data subsumed by it, in light of current research. *Am. Nat., 72,* 416–433.

Huxley, J. S. (1938b). The present standing of the theory of sexual selection. In G. R. de Beer, ed., *Evolution: Essays on Aspects of Evolutionary Biology,* pp. 11–42. Oxford: Clarendon Press.

Iwahashi, O., and Majima, T. (1986). Lek formation and male-male competition in the melon fly *Dacus cucurbitae* (Diptera: Tephritidae). *Appl. Entomol. Zool., 21*, 70–75.

Iwasa, Y., Pomiankowski, A., and Nee, S. (1991). The evolution of costly mate preferences. II. The "handicap" principle. *Evolution, 45*, 1431–1442.

Jamieson, I. G., and Craig, J. L. (1987). Dominance and mating in a communal polygynandrous bird: Cooperation and indifference towards mating competitors? *Ethology, 75*, 317–327.

Janetos, A. C. (1980). Strategies of female choice: A theoretical analysis. *Behav. Ecol. Sociobiol., 7*, 107–112.

Jehl, J. R., and Murray, B. G., Jr. (1986). The evolution of normal and reverse sexual size dimorphism in shorebirds and other birds. In R. F. Johnston, ed., *Current Ornithology*, pp. 1–86. New York: Plenum Press.

Johnson, L. L., and Boyce, M. S. (1991). Female choice of males with low parasite loads in sage grouse. In J. E. Loye and M. Zuk, eds., *Bird-Parasite Interaction: Ecology, Evolution and Behaviour*, pp. 377–388. Oxford: Oxford University Press.

Keenleyside, M.H.A. (1991). *Cichlid Fishes: Behaviour, Ecology and Evolution*. London: Chapman and Hall.

Kennedy, C.E.J., Endler, J. A., Poynton, S. L., and McMinn, H. (1987). Parasite load predicts mate choice in guppies. *Behav. Ecol. Sociobiol., 21*, 291–295.

Kenward, R. E. (1978). Hawks and doves: Attack success and selection in goshawk flights at wood-pigeons. *J. Anim. Ecol., 47*, 449–460.

Kimsey, L. S. (1980). The behaviour of male orchid bees (Apidae, Hymenoptera, Insecta) and the question of leks. *Anim. Behav., 28*, 996–1004.

Kirkpatrick, M. (1982). Sexual selection and the evolution of female choice. *Evolution, 36*, 1–12.

Kirkpatrick, M. (1985). Evolution of female choice and male parental investment in polygamous species: The demise of the sexy son. *Am. Nat., 125*, 788–810.

Kirkpatrick, M. (1986). The handicap mechanism of sexual selection does not work. *Am. Nat., 127*, 222–240.

Kirkpatrick, M. (1987). Sexual selection by female choice in polygynous animals. *Ann. Rev. Ecol. Syst., 18*, 43–70.

Kirkpatrick, M., and Ryan, M. J. (1991). The evolution of mating preferences and the paradox of the lek. *Nature (London), 350*, 33–38.

Kirschofer, R. (1953). Aktionssystem des Maulbrüters *Haplochromis desfontainesii. Z. Tierpsychol., 10*, 297–318.

Klump, G. M., and Gerhardt, H. C. (1987). Use of non-arbitrary acoustic criteria in mate choice by female grey treefrogs. *Nature (London), 326*, 286–288.

Knapton, R. W. (1985). Lek structure and territoriality in the chryxus arctic butterfly, *Oeneis chryxus* (Satyridae). *Behav. Ecol. Sociobiol., 17*, 389–395.

Kodrick-Brown, A. (1977). Reproductive success and the evolution of breeding territories in pupfish (Cyprinodon). *Evolution, 31*, 750–766.

Kodrick-Brown, A., and Brown, J. H. (1984). Truth in advertising: The kinds of traits favoured by sexual selection. *Am. Nat., 124*, 309–323.

Köhler, D., and Tembrock, G. (1987). Acoustic signals in a Wolf spider *Hygrolucosa rubrofasciata* (Arachnida: Lycosidae). *Zool. Anz., 219*, 147–153.

Koivisto, I. (1965). Behavior of the black grouse, *Lyrurus tetrix* (L.), during the spring display. *Finn. Game Res., 26,* 5–60.

Krebs, J. R., and Davies, N. B. (1993). *An Introduction to Behavioural Ecology.* 3d ed. Oxford: Blackwell.

Krebs, J. R., and Harvey, P. H. (1988). Lekking in Florence. *Nature (London), 333,* 12–13.

Kruijt, J. P., and de Vos, G. J. (1988). Individual variation in reproductive success in male black grouse, *Tetrao tetrix* L. In T. H. Clutton-Brock, ed., *Reproductive Success. Studies of Individual Variation in Contrasting Breeding Systems,* pp. 279–290. Chicago: University of Chicago Press.

Kruijt, J. P., de Vos, G. J., and Bossema, I. (1972). The arena system of the black grouse. *Proc. 15th Int. Ornith. Congr.,* 399–423.

Kruijt, J. P., and Hogan, J. A. (1967). Social behaviour on the lek in black grouse *Lyrurus tetrix tetrix* (L.). *Ardea, 55,* 203–240.

Lack, D. (1968). *Ecological Adaptations for Breeding in Birds.* London: Chapman and Hall.

Landau, H. G. (1951). On dominance relations and the structure of animal societies. 1. Effect of inherent characteristics. *Bull. Math. Biophys., 13,* 1–19.

Lande, R. (1980). Sexual dimorphism, sexual selection, and adaptation in polygenic characters. *Evolution, 34,* 292–305.

Lande, R. (1981). Models of speciation by sexual selection on polygenic characters. *Proc. Natl. Acad. Sci. USA, 78,* 3721–3725.

Lande, R. (1987). Genetic correlations between the sexes in the evolution of sexual dimorphism and mating preferences. In J. W. Bradbury and M. B. Andersson, eds., *Sexual Selection: Testing the Alternatives,* pp. 83–94. Chichester, U.K.: John Wiley.

Landel, H. In prep. Correlates of male mating success in the sharp-tailed grouse, *Tympanuchus phasianellus.*

Langbein, J., and Thirgood, S. J. (1989). Variation in mating systems of fallow deer *(Dama dama)* in relation to ecology. *Ethology, 83,* 195–214.

Lank, D. B., and Smith, C. M. (1987). Conditional lekking in the ruff (*Philomachus pugnax*). *Behav. Ecol. Sociobiol., 20,* 137–145.

Lank, D. B., and Smith, C. M. (1992). Females prefer larger leks: Field experiments with ruffs (*Philomachus pugnax*). *Behav. Ecol. Sociobiol., 30,* 323–329.

Larsson, F. (1989) Mating patterns in six insect species: Effects of weather and population density. Ph.D. thesis, Uppsala University.

Lazenby-Cohen, K. A., and Cockburn, A. (1988). Lek promiscuity in a semelparous mammal *Antechinus stuartii* (Marsupialia: Dasyuridae)? *Behav. Ecol. Sociobiol., 22,* 195–202.

Leader-Williams, N. (1988). *Reindeer on South Georgia.* Cambridge, U.K.: Cambridge University Press.

Le Croy, M. (1981). The genus *Paradisaea*—Display and evolution. *Am. Mus. Nov., 2714,* 1–52.

Lederhouse, R. C. (1982). Territorial defense and lek behaviour of the black swallowtail butterfly, *Papilio polyxenes. Behav. Ecol. Sociobiol., 10,* 109–118.

Lemnell, P. A. (1978). Social behaviour of the great snipe *Capella media* at the arena display. *Ornis Scand., 9,* 146–165.

Lewis, R. A. (1985). Do blue grouse form leks? *Auk, 102*, 180–184.

Lill, A. (1974). Sexual behavior of the lek-forming white-bearded manakin *(Manacus manacus trinitatis* Hartert). *Z. Tierpsychol., 36*, 1–36.

Lill, A. (1976). Lek behavior in the golden-headed manakin, *Pipra erythrocephala* in Trinidad (West Indies). *Z. Tierpsychol. Suppl., 18*, 1–84.

Lloyd, J. E. (1972). Mating behavior of a New Guinea Luciola firefly: A new communicative protocol. *Coleop. Bull., 26*, 155–163.

Lloyd, J. E. (1979). Sexual selection in luminescent beetles. In M. S. Blum and N. A. Blum, eds., *Sexual Selection and Reproductive Competition in Insects*, pp. 31–76. New York: Academic Press.

Lloyd, L. (1867). *Game Birds and Wild Fowl of Sweden and Norway*. London: Frederick Warne.

Løfaldli, L., Kålås, J. A., and Fiske, P. (1992). Habitat selection and diet of great snipe *Gallinago media* during breeding. *Ibis, 134*, 35–43.

Loiselle, P. V., and Barlow, G. W. (1978). Do fishes lek like birds? In E. S. Reese and F. J. Lighter, eds., *Contrasts in Behavior*, pp. 31–76. Chichester, U.K.: John Wiley.

Losey, G. S., Jr., Stanton, F. G., Telecky, T. M., Tyler, W. A., III, and the Zoology 691 Graduate Seminar Class (1986). Copying others, an evolutionarily stable strategy for mate choice: A model. *Am. Nat., 128*, 653–664.

Lott, D. F. (1991). *Intraspecific Variation in the Social Systems of Wild Vertebrates*. Cambridge, U.K.: Cambridge University Press.

Lowe, R. H. (1952). Report on the Tilapia and other fish and fisheries of Lake Nyasa, 1945–47. *Colonial Off. Fish. Publ., 1*, 1–26.

Lowe-McConnell, R. H. (1956). The breeding behaviour of *Tilapia* species (Pisces: Cichlidae) in natural waters: Observations on *T. karomo* and *T. variabilis* Boulenger. *Behaviour, 9*, 140–163.

Lowe-McConnell, R. H. (1957). Observations on the diagnosis and biology of *Tilapia leucosticta* Trewavas in East Africa. *Rev. Zool. Bot. Afr., 55*, 353–373.

Lowe-McConnell, R. H. (1958). Observations on the biology of *Tilapia nilotica* Linné in East African waters. *Rev. Zool. Bot. Afr., 57*, 129–170.

Lowe-McConnell, R. H. (1959). Breeding behaviour patterns and ecological differences between *Tilapia* species and their significance for the evolution within the genus *Tilapia* (Pisces: Cichlidae). *Proc. Roy. Soc. London B, 132*, 1–30.

Loye, J. E., and Zuk, M., eds. (1991). *Bird-Parasite Interactions: Ecology, Evolution and Behaviour*. Oxford: Oxford University Press.

Ludwig, A. (1894). *Das Birkwild*. Vienna.

Lumsden, H. G. (1961). The display of the capercaillie. *Brit. Birds, 54*, 257–272.

Lumsden, H. G. (1965). Displays of the sharp-tailed grouse. *Ont. Dep. Lands For. Res. Rep., 66*, 1–68.

Lund, H. M.-K. (1946). Entoparasites in the capercailzie (Tetrao urogallus). *Skandinav. Veterinärtidskr.*, 641–662.

Lund, H. M.-K. (1954). *Nematodes, Cestodes and Coccidia Found in 136 Black Grouse (Lyrurus tetrix) in Norway*. Oslo: Statens Viltundersøkelser.

Lyon, B., and Montgomerie, R. (1986). Delayed plumage maturation in passerine birds: Reliable signalling by subordinate males? *Evolution, 40*, 605–615.

Manning, J. T. (1989). Age-advertisement and the evolution of the peacock's train. *J. Evol. Biol., 2,* 379–384.

Marius-Jestin, V., Le Menec, M., Thibault, E., Moisan, J. C., and L'Hospitalier, R. (1987). Normal phallus flora of the gander. *J. Vet. Med. B., 34,* 67–78.

Maynard Smith, J. (1958). Sexual selection. In S. A. Barnett, ed., *A Century of Darwin,* pp. 231–244. London: Heinemann.

Maynard Smith, J. (1976). Sexual selection and the handicap principle. *J. theor. Biol., 57,* 239–242.

Maynard Smith, J. (1983). *Evolution and the Theory of Games.* Cambridge, U.K.: Cambridge University Press.

Maynard Smith, J. (1991). Theories of sexual selection. *TREE, 6,* 146–151.

Maynard Smith, J., and Price, G. R. (1973). The logic of animal conflict. *Nature (London), 246,* 15–18.

McCullagh, P., and Nelder, J. A. (1989). *Generalized Linear Models.* 2d ed. London: Chapman and Hall.

McDonald, D. B. (1989). Cooperation under sexual selection: Age-graded changes in a lekking bird. *Am. Nat., 134,* 709–730.

McDonald, D. B. (1990). Correlates of male mating success in a lekking bird with male-male cooperation. *Anim. Behav., 37,* 1007–1022.

McKaye, K. R. (1983). Ecology and breeding behavior of a cichlid fish, *Cyrtocara eucinostomus,* on a large lek in Lake Malawi, Africa. *Env. Biol. Fish., 8,* 81–96.

McKaye, K. R., Louda, S. M., and Stauffer, J. R., Jr. (1990). Bower size and male reproductive success in a cichlid fish lek. *Am. Nat., 135,* 597–613.

McKitrick, M. C. (1992). Phylogenetic analysis of avian parental care. *Auk, 109,* 828–846.

Medawar, P. B. (1952). *An Unsolved Problem of Biology.* London: H. K. Lewis.

Merton, D. V., Morris, R. V., and Atkinson, I.A.E. (1984). Lek behaviour in a parrot: The kakapo, *Strigops habroptilus,* of New Zealand. *Ibis, 126,* 277–283.

Michod, R. E., and Hasson, O. (1990). On the evolution of reliable indicators of fitness. *Am. Nat., 135,* 788–808.

Milinski, M., and Bakker, T.C.M. (1990). Female sticklebacks use male coloration in mate choice and hence avoid parasitized males. *Nature (London), 344,* 330–333.

Milinski, M., and Parker, G. A. (1991). Competition for resources. In J. R. Krebs and N. B. Davies, eds., *Behavioural Ecology: An Evolutionary Approach,* pp. 137–168. Oxford: Blackwell.

Mjelstad, H. (1991). Displaying intensity and sperm quality in the capercaillie *Tetrao urogallus. Fauna Norv. Ser. C, Cinclus, 14,* 93–94.

Møller, A. P. (1990). Effects of a haematophagus mite on the barn swallow (*Hirundo rustica*): A test of the Hamilton and Zuk hypothesis. *Evolution, 44,* 771–784.

Møller, A. P. (1991). Viability is positively related to degree of ornamentation in male swallows. *Proc. Roy. Soc. London B, 41,* 145–148.

Møller, A. P., and Birkhead, T. R. (1991). Frequent copulations and mate guarding as alternative paternity guards in birds: A comparative study. *Behaviour, 118,* 170–186.

Møller, P. P., and Pomiankowski, A. (1993). Why have birds got multiple sexual ornaments? *Behav. Ecol. Sociobiol., 32*, 167–176.

Morgante, J. S., Malavasi, A., and Prokopy, R. J. (1983). Mating behavior of wild *Anastrepha fraterculus* (Diptera: Tephritidae) on a caged host tree. *Fla. Entomol., 66*, 234–241.

Morris, G. K., Kerr, G. E., and Fullard, J. H. (1978). Phonotactic preferences of female meadow katydids (Orthoptera: Tettigoniidae: *Conocephalus nigropleurum*). *Can. J. Zool., 53*, 1127–1130.

Moyer, J. T., and Yogo, Y. (1982). The lek-like mating system of *Halichoeres melanochir* (Pisces: Labridae) at Miyoke-juma, Japan. *Z. Tierpsychol., 60*, 209–226.

Myers, J. P. (1979). Leks, sex and buff-breasted sandpipers. *Am. Birds, 33*, 823–825.

Neems, R. M., Lazarus, J., and Mclachlan, A. J. (1992). Swarming behavior in male chironomid midges: A cost benefit analysis. *Behav. Ecol., 3*, 285–290.

Nichols, R. A., and Butlin, R. K. (1989). Does runaway sexual selection work in finite populations? *J. Evol. Biol., 2*, 299–313.

Nilsson, S. (1824). *Skandinavisk fauna, foglarna*. Lund: Berlingska boktryckeriet.

Nishida, T. (In press). Spatial relationships between mate acquisition probability and aggregation size in a gregarious coreid bug *(Colpula lativentris)*: A case of the ideal free distribution under perceptual constraints. *Behav. Ecol.*.

Nitchuk, W. M., and Evans, R. M. (1978). A volumetric analysis of sharp-tailed grouse sperm in relation to dancing ground size and organization. *Wilson Bull., 90*, 460–462.

Nur, N., and Hasson, O. (1984). Phenotypic plasticity and the handicap principle. *J. theor. Biol., 110*, 275–297.

Oakes, E. J. (1992). Lekking and the evolution of sexual dimorphism in birds: Comparative approaches. *Am. Nat., 140*, 665–684.

O'Donald, P. (1962). The theory of sexual selection. *Heredity, 17*, 541–552.

O'Donald, P. (1980). *Genetic Models of Sexual Selection*. Cambridge, U.K.: Cambridge University Press.

Orians, G. H. (1969). On the evolution of mating systems in birds and mammals. *Am. Nat., 103*, 589–603.

Oring, L. W. (1982). Avian mating systems. In D. S. Farner, J. R. King, and K. C. Parkes, eds., *Avian Biology*, pp. 1–92. New York: Academic Press.

Otronen, M. (1988). Intra—and intersexual interactions at breeding burrows in the horned beetle, *Coprophanaeus ensifer*. *Anim. Behav., 36*, 741–748.

Otronen, M. (1990). Mating behaviour and sperm competition in the dung fly *Dryomyza anilis*. *Behav. Ecol. Sociobiol., 26*, 349–356.

◆ Otte, D. (1974). Effects and functions in the evolution of signalling systems. *Ann. Rev. Ecol. Syst., 5*, 385–417.

Owens, I.P.F., Burke, T., Thompson, D.B.A. (1994). Extraordinary sex roles in the Eurasian dotterel: Female mating arenas, female-female competition, and female mate choice. *Am. Nat., 144*, 76–100.

Pagel, M. D. (1992). A method for the analysis of comparative data. *J. theor. Biol., 156*, 431–442.

Pagel, M. D., and Harvey, P. H. (1988). Recent developments in the analysis of comparative data. *Q. Rev. Biol., 63*, 413–440.

Palokangas, P., Alatalo, R. V., and Korpimäki, E. (1992). Female choice in the kestrel under different availability of mating options. *Anim. Behav., 43*, 659–665.

Parker, G. A. (1970). The reproductive behaviour and the nature of sexual selection in *Scatophaga stercoraria* L. (Diptera: Scatophagidae). II. The fertilization rate and the spatial and temporal relationships of each sex around the site of mating and oviposition. *J. Anim. Ecol., 39*, 205–229.

Parker, G. A. (1978a). Evolution of competitive mate searching. *Ann. Rev. Entomol., 23*, 173–196.

Parker, G. A. (1978b). Searching for mates. In J. R. Krebs and N. B. Davies, eds., *Behavioural Ecology: An Evolutionary Approach*, pp. 214–244. Oxford: Blackwell.

Parker, G. A. (1982). Phenotype-limited evolutionary strategies. In K.C.S. Group, ed., *Current Problems in Sociobiology*, pp. 173–201. Cambridge, U.K.: Cambridge University Press.

Parker, G. A. (1983). Mate quality and mating decisions. In P. Bateson, ed., *Mate Choice*, pp. 141–166. Cambridge, U.K.: Cambridge University Press.

Parker, G. A. (1984a). Evolutionary stable strategies. In J. R. Krebs and N. B. Davies, eds., *Behavioural Ecology: An Evolutionary Approach*, pp. 30–61. Oxford: Blackwell.

Parker, G. A. (1984b). The producer/scrounger model and its relevance to sexuality. In C. J. Barnard, ed., *Producers and Scroungers: Strategies of Exploitation and Parasitism*, pp. 127–183. Beckenham, U.K.: Croom Helm.

Parker, G. A., and Sutherland, W. J. (1986). Ideal free distributions when individuals differ in competitive ability: Phenotype-limited ideal free models. *Anim. Behav., 34*, 1222–1242.

Parker, H., Mjelstad, H., and Solheim, J. T. (1989). Duration of fertility in capercaillie hens after separation from males. *Ornis Scand., 20*, 307–310.

Parsons, P. A. (1977a). Lek behavior in *Drosophila (Hirtodrosophila) polypori* Malloch—an Australian rain forest species. *Evolution, 31*, 223–225.

Parsons, P. A. (1977b). Lek behavior in three species of the subgenus *Hirtodrosophila* of Australian *Drosophila*. *Nature (London), 265*, 48.

Parsons, P. A. (1978). Habitat selection and evolutionary strategies in *Drosophila:* an invited address. *Behav. Gen., 8*, 511–526.

Partridge, L., and Halliday, T. (1984). Mating patterns and mate choice. In J. R. Krebs and N. B. Davies, eds., *Behavioural Ecology: An Evolutionary Approach*, pp. 222–250. Oxford: Blackwell.

Patterson, I. J. (1977). Aggression and dominance in winter flocks of shelduck *Tadorna tadorna* (L.). *Anim. Behav., 25*, 447–459.

Payne, R. B. (1973). Behavior, mimetic songs and song dialects, and the relationship of the parasitic indigobirds (*Vidua*) of Africa. *Ornith. Monogr., 11*, 1–333.

Payne, R. B. (1984). Sexual selection, lek and arena behavior, and sexual size dimorphism in birds. *Orn. Monogr., 33*, 1–52.

Payne, R. B., and Payne, K. (1977). Social organization and mating success in local song populations of village indigobirds, *Vidua chalybeata*. *Z. Tierpsychol., 45*, 113–173.

Pemberton, J. M., and Balmford, A. (1987). Lekking in fallow deer. *J. Zool., 213*, 762–765.

Perrill, S. A., Gerhardt, H. C., and Daniel, R. (1978). Sexual parasitism in the green treefrog, *Hyla cinerea. Science, 200*, 1179–1180.

Petersson, E. (1989). Mating in swarming caddis flies (*Trichoptera: Leptoceridae*). Ph.D. thesis, Uppsala University.

Petrie, M. (1992). Peacocks with low mating success are more likely to suffer predation. *Anim. Behav., 44*, 585–586.

Petrie, M. (In press). The offspring of peacocks with more elaborate trains show better growth and survival. *Nature* (Lond.).

Petrie, M., Hall, M., Halliday, T., Budgey, H., and Pierpoint, C. (1992). Multiple mating in a lekking bird: Why do peheans mate with more than one male and the same male more than once? *Behav. Ecol. Sociobiol., 31*, 349–358.

Petrie, M., Halliday, T., and Sanders, C. (1991). Peahens prefer peacocks with elaborate trains. *Anim. Behav., 41*, 323–332.

Petrie, M., and Williams, A. (1993). Peahens lay more eggs for peacocks with larger trains. *Proc. Roy. Soc. London B, 251*, 127–131.

Phillips, J. B. (1990). Lek behaviour in birds: Do displaying males reduce nest predation? *Anim. Behav., 39*, 555–565.

Pomiankowski, A. (1987a). Sexual selection: The handicap principle does work—sometimes. *Proc. Roy. Soc. London B, 231*, 123–145.

Pomiankowski, A. (1987b). The 'handicap principle' works without Fisher. *TREE, 2*, 2–3.

Pomiankowski, A. (1988). The evolution of female mate preferences for male genetic quality. *Oxford Surv. Evol. Biol., 5*, 136–184.

Pomiankowski, A. (1990). How to find the top male. *Nature (London), 347*, 616–617.

Pomiankowski, A., Iwasa, Y., and Nee, S. (1991). The evolution of costly mate preferences. I. Fisher and biased mutation. *Evolution, 45*, 1422–1430.

Porter, W. F. (1985). Turkey. In C. B. and E. Lack, eds., *A Dictionary of Birds*, pp. 613–614. Calton: T. and A. D. Poyser.

Price, T., Schluter, D., and Heckman, N. E. (1993). Sexual selection when the female directly benefits. *Biol. J. Linn. Soc., 48*, 187–211.

Pruett-Jones, M., and Pruett-Jones, S. G. (1982). Spacing and distribution of bowers in MacGregor's bowerbird (*Amblyornis macgreogoriae*). *Behav. Ecol. Sociobiol., 11*, 25–32.

Pruett-Jones, S. G. (1988). Lekking versus solitary display: Temporal variations in dispersion in the buff-breasted sandpiper. *Anim. Behav., 36*, 1740–1752.

Pruett-Jones, S. (1992). Independent versus non-independent mate choice: Do females copy each other? *Am. Nat., 140*, 1000–1009.

Pruett-Jones, S. G., and Pruett-Jones, M. A. (1991). Sexual selection through female choice in Lawe's parotia, a lek-mating bird of paradise. *Evolution, 44*, 486–501.

Prum, R. O. (1990). Phylogenetic analysis of the evolution of display behavior in the neotropical manakins (Aves: Pipridae). *Ethology, 84*, 202–231.

Prum, R. O. (1992). Syringeal morphology, phylogeny, and evolution of the neotropical manakins (Aves: Pipridae). *Amer. Mus. Nat. Hist. Novitates, 3043*.

Prum, R. O. (In press). Phylogenetic analysis of the evolution of alternative social behavior in the manakins (Aves: Pipridae). *Evolution.*

Pukowski, E. (1933). Ökologische Untersuchungen an Necrophorus F. *Z. Morph. Ökol. Tiere, 27,* 518–586.

● Pulliam, H. R. (1973). On the advantages of flocking. *J. theor. Biol., 38,* 419–422.

Pulliam, H. R., and Caraco, T. (1984). Living in groups: Is there an optimal group size? In J. R. Krebs and N. B. Davies, eds., *Behavioural Ecology: An evolutionary approach,* pp. 122–147. Oxford: Blackwell.

Purvis, A. (1991). *Comparative Analysis by Independent Contrasts, Version 1.2: User's Guide.* Oxford University.

Queller, D. C. (1987). The evolution of leks through female choice. *Anim. Behav., 35,* 1424–1432.

Ralls, K. (1976). Mammals in which females are larger than males. *Q. Rev. Biol., 51,* 245–276.

Ranger, G. A. (1955). On the three species of honeyguide: The greater *(Indicator indicator),* the lesser *(I. minor)* and the scaly-throated *(I. variegatus). Ostrich, 26,* 70–87.

Rauch, N. (1985). Female habitat choice as a determinant of the reproductive success of the territorial male marine iguana *(Amblyrhynchus cristatus). Behav. Ecol. Sociobiol., 16,* 125–134.

Read, A. (1988). Sexual selection and the role of parasites. *TREE, 3,* 97–102.

Real, L. (1990). Search theory and mate choice. I. Models of single sex discrimination. *Am. Nat., 136,* 376–405.

Reynolds, J. D., and Gross, M. R. (1990). Costs and benefits of female mate choice: Is there a lek paradox? *Am. Nat., 136,* 230–243.

Ricklefs, R. (1969). An analysis of nesting mortality in birds. *Smithson. Contr. Zool., 9,* 1–48.

Ridley, M. (1983). *The Explanation of Organic Diversity: The Comparative Method and Adaptations for Mating.* Oxford: Oxford University Press.

Ringo, J. M. (1976). A communal display in Hawaiian *Drosophila* (Diptera: Drosophilidae). *Ann. Entomol. Soc. Am., 69,* 209–214.

Rintamäki, P. T., Alatalo, R. V., Höglund, J., and Lundberg, A. In press. Male territoriality and female choice on black grouse leks. *Anim. Behav.*

Rintamäki, P. T., Alatalo, R. V., Höglund, J., and Lundberg, A. In prep. Mate sampling behaviour of black grouse females *(Tetrao tetrix).*

Rippin, A. B., and Boag, D. A. (1974). Spatial organization among sharp-tailed grouse on arenas. *Can. J. Zool., 52,* 591–597.

Robbins, M. B. (1983). The display repertoire of the band-tailed manakin *(Pipra fasciicauda). Wilson Bull., 95,* 321–504.

Robbins, M. B. (1985). Social organization of the band-tailed manakin *(Pipra fasciicauda). Condor, 87,* 449–456.

Robel, R. J. (1970). Possible role of behavior in regulating greater prairie chicken populations. *J. Wildl. Manag., 34,* 306–312.

Robel, R. J., and Ballard, W. B., Jr. (1974). Lek social organization and reproductive success in the greater prairie chicken. *Am. Zool., 14,* 121–128.

Robertson, H. M. (1985). Female dimorphism and mating behaviour in a damselfly, *Ischnura ramburi:* Females mimicking males. *Anim. Behav., 33,* 805–809.

Robertson, J.G.M. (1984). Acoustic spacing by breeding males of *Uperoleia rugosa* (Anura: Leptodactylidae). *Z. Tierpsychol., 64*, 283–297.

Robertson, J.G.M. (1986a). Female choice, male strategies and the role of vocalizations in the Australian frog *Uperoleia rugosa. Anim. Behav., 34*, 773–784.

Robertson, J.G.M. (1986b). Male territoriality, fighting and assessment of fighting ability in the Australian frog *Uperoleia rugosa. Anim. Behav., 34*, 763–772.

Robertson, J.G.M. (1990). Female choice increases fertilization success in the Australian frog, *Uperoleia laevigata. Anim. Behav., 39*, 639–645.

Robinette, W. L., and Child, G.F.T. (1964). Notes on biology of the lechwe (*Kobus leche*). *Puku, 2*, 84–117.

Ruwet, J. C. (1962). La reproduction des *Tilapia machrochir* Blgr. et *T. melanopleura* Dum. au lac de la retenue de la Lufira (Katanga). *Rev. Zool. Bot. Afr., 66*, 244–271.

Ryan, M. J. (1980a). Female mate choice in a Neotropical frog. *Science, 209*, 523–525.

Ryan, M. J. (1980b). The reproductive behavior of the bullfrog (*Rana catesbeiana*). *Copeia, 1980*, 108–114.

Ryan, M. J. (1983). Sexual selection and communication in a Neotropical frog, *Physalaemus pustulosus. Evolution, 37*, 261–272.

Ryan, M. J. (1985). *The Tùngara Frog: A Study in Sexual Selection and Communication.* Chicago: University of Chicago Press.

Ryan, M. J. (1991a). Sexual selection and communication in frogs. *TREE, 6*, 351–355.

Ryan, M. J. (1991b). Sexual selection, sensory systems and sensory exploitation. *Oxford Surv. Evol. Biol., 7*, 157–195.

Ryan, M. J., Fox, J. H., Wilczynski, W., and Rand, A. S. (1990). Sexual selection by sensory exploitation in the frog *Physalaemus pustulosus. Nature (London), 343*, 66–67.

Ryan, M. J., and Rand, A. S. (1990). The sensory basis of sexual selection for complex calls in the tùngara frog, *Physalaemus pustulosus* (sexual selection for sensory exploitation). *Evolution, 44*, 305–314.

Ryan, M. J., Tuttle, M. D., and Rand, A. S. (1982). Bat predation and sexual advertisement in a neotropical frog. *Am. Nat., 119*, 136–139.

Ryan, M. J., Tuttle, M. D., and Taft, L. K. (1981). The costs and benefits of frog chorusing behavior. *Behav. Ecol. Sociobiol., 8*, 273–278.

Sæther, B.-E., and Andersen, R. (1988). Ecological consequences of body size in grouse *Tetraonidae. Fauna Norv. Ser C, Cinclus, 11*, 19–26.

Sakaluk, S. K., and Belwood, J. J. (1984). Gecko phonotaxis to cricket calling: A case of satellite predation. *Anim. Behav., 32*, 659–662.

Schaal, A., and Bradbury, J. W. (1987). Lek breeding in a deer species. *Biol. Behav., 12*, 28–32.

Schroeder, M. A., and White, G. C. (1993). Dispersion of greater prairie chicken nests in relation to lek location: Evaluation of the hotspot hypothesis of lek evolution. *Behav. Ecol., 4*, 266–270.

Schwartz, P., and Snow, D. W. (1978). Display and related behaviour of the wire-tailed manakin. *Living Bird, 17*, 51–78.

Scott, J. W. (1942). Mating behaviour of the sage grouse. *Auk, 59*, 477–498.

Selander, R. K. (1972). Sexual selection and dimorphism in birds. In B. Campbell, eds., *Sexual Selection and the Descent of Man*, pp. 180–230. Chicago: Aldine.

Selous, E. (1906). Observations tending to throw light on the questions of sexual selection in birds, including a day-to-day diary on the breeding habits of the ruff *(Machetes pugnax)*. *Zoologist, 10*, 201–219, 285–294, 419–428.

Selous, E. (1907). Observations tending to throw light on the questions of sexual selection in birds, including a day-to-day diary on the breeding habits of the ruff *(Machetes pugnax)*. *Zoologist, 11*, 60–65, 161–182, 367–381.

Selous, E. (1927). *Realities of Bird Life*. London: Constable.

Shaw, K. C., North, R. C., and Meixner, A. J. (1981). Movement and spacing of singing *Amblyocorpha parvipennis* males (Tettigonidae: Phaneropterinae). *Ann. Entomol. Soc. Am., 74*, 436–444.

Shaw, P. (1984). The social behavior of the pin-tailed whydah *Vidua macroura* in northern Ghana. *Ibis, 126*, 463–473.

Sheldon, B. C. (1993). Sexually transmitted disease in birds: Occurrence and evolutionary significance. *Phil. Trans. Roy. Soc. London B, 339*, 491–497.

Shelly, T. E. (1987). Lek behaviour of a Hawaiian *Drosophila:* Male spacing, aggression and female visitation. *Anim. Behav., 35*, 1394–1404.

Shelly, T. E. (1988). Lek behaviour of *Drosophila cnecopleura* in Hawaii. *Ecol. Entomol., 13*, 51–55.

Shelly, T. E. (1990). Waiting for mates: Variation in female encounter rates within and between leks of *Drosophila conformis. Behaviour, 107*, 34–48.

Shelly, T. E., and Greenfield, M. D. (1985). Alternative mating strategies in a desert grasshopper: A transitional analysis. *Anim. Behav., 33*, 1211–1222.

Shelly, T., and Kaneshiro, K. Y. (1991). Lek behaviour of the oriental fruit fly, *Dacus dorsalis*, in Hawaii (Diptera: Tephritidae). *J. Insect Behav., 4*, 235–241.

Shepard, J. M. (1976). Factors influencing female choice in the lek-mating system of the ruff. *Living Bird, 14*, 87–111.

Shields, O. (1967). Hilltopping. *J. Res. Lepid., 6*, 69–178.

Sibley, C. G., and Ahlquist, J. E. (1990). *Phylogeny and Classification of Birds: A Study in Molecular Evolution*. New Haven: Yale University Press.

Sibly, R. (1984). Models of producer/scrounger relationships between and within species. In C. J. Barnard, ed., *The Producer/Scrounger Model and Its Relevance to Sexuality*, pp. 267–287. Beckenham, U.K.: Croom Helm.

Sick, H. (1967). Courtship behavior of the manakins (Pipridae): A review. *Living Bird, 6*, 5–22.

Skutch, A. F. (1969). *Life Histories of Central American Birds, III*. Pacific Coast Avifauna, no. 35.

Snow, B. K. (1973). The behaviour and ecology of hermit hummingbirds in the Kanaku mountains, Guyana. *Wilson Bull., 85*, 163–177.

Snow, B. K. (1974). Lek behaviour and breeding of Guy's hermit hummingbird *Phaethornis guy. Ibis, 116*, 278–297.

Snow, B. K., and Snow, D. W. (1979). The ochre-bellied flycatcher and the evolution of lek behaviour. *Condor, 81*, 286–292.

Snow, D. W. (1962). A field study of the black and white manakin, *Manacus manacus*, in Trinidad. *Zoologica, 47*, 65–104.

Snow, D. W. (1963a). The evolution of manakin displays. *Proc. 13th Int. Ornith. Congr.*, 553–561.

Snow, D. W. (1963b). The display of the orange-headed manakin. *Condor, 65*, 44–48.

Snow, D. W. (1963c). The display of the blue-backed manakin, *Chiroxiphia pareola,* in Tobago, W.I. *Zoologica, 48*, 167–176.

Snow, D. W. (1982). *The Cotingas.* Oxford: Oxford University Press.

Soper, R. S., Shewell, G. E., and Tyrrell, D. (1976). *Colcondamyia auditrix* nov. sp. (Diptera: Sarcophagidae), a parasite which is attracted by the mating song of its host, *Okonagana rumorsa* (Homoptera: Cicadidae). *Can. Entomol., 108*, 61–68.

Spieth, H. T. (1978). Courtship patterns and evolution of the *Drosophila adiastola* and *planitibia* species subgroups. *Evolution, 32*, 435–451.

Spurrier, M. F., Boyce, M. S., and Manly, B.F.J. (1991). Effects of parasites on mate choice by captive sage grouse. In J. E. Loye and M. Zuk, eds., *Bird-Parasite Interactions*, pp. 389–398. Oxford: Oxford University Press.

Stamps, J. A. (1977). Social behavior and spacing patterns in lizards. In C. Gans and D. W. Tinkle, eds., *Biology of the Reptilia*, pp. 265–334. New York: Academic Press.

Steiner, A. L. (1978). Observations on spacing, aggressive, and lekking behavior of digger wasp males of *Eucerceris flavocincta* (Hymenoptera: Sphecidae; Cercerini). *J. Kans. Entomol. Soc., 51*, 492–498.

Stiles, F. G. (1973). Food supply and the annual cycle of the Anna hummingbird. *Univ. Calif. Publ. Zool., 97*, 1–109.

Stiles, F. G., and Whitney, B. (1983). Notes on the behavior of the Costa Rican Sharpbill (*Oxyruncus cristatus frater*). *Auk, 100*, 117–125.

Stiles, F. G., and Wolf, L. L. (1979). Ecology and evolution of lek mating behaviour in the long-tailed hermit hummingbird. *Ornith. Monogr., 27*, 1–77.

• Stillman, R., Clutton-Brock, T. H., and Sutherland, W. J. (1993). Black holes, mate retention and the evolution of ungulate leks. *Behav. Ecol., 4*, 1–6.

Stipkovits, L., Varga, Z., Czifra, G., and Dobos-Kovacs, M. (1986). Occurrence of mycoplasmas in geese infected with inflammation of the cloaca and phallus. *Avian Pathol., 15*, 289–299.

Sullivan, B. K. (1982). Sexual selection in Woodhouse's toad *(Bufo woodhousei)*. I. Chorus organization. *Anim. Behav., 30*, 680–686.

Sullivan, B. K. (1983). Sexual selection in Woodhouse's toad *(Bufo woodhousei)*. II. Female choice. *Anim. Behav., 30*, 680–686.

Sullivan, B. K., and Hinshaw, S. H. (1992). Female choice and selection of male calling behaviour in the grey treefrog *Hyla versicolor. Anim. Behav., 44*, 733–744.

Sullivan, R. T. (1981). Insect swarming and mating. *Fla. Entomol., 64*, 44–65.

Sutherland, W. J. (1983). Aggregation and the ideal free distribution. *J. Anim. Ecol., 52*, 821–828.

Sutherland, W. J. (1985a). Chance can produce a sex difference in variance in mating success and account for Bateman's data. *Anim. Behav., 33*, 1349–1352.

Sutherland, W. J. (1985b). Measures of sexual selection. *Oxford Surv. Evol. Biol., 2*, 90–101.

Sutherland, W. J. (1987). Random and deterministic components of variance in mating success. In J. W. Bradbury and M. B. Andersson, eds., *Sexual Selection: Testing the Alternatives*, pp. 209–219. Chichester, U.K.: John Wiley.

Sutherland, W. J., Höglund, J., and Widemo, F. (Subm.). Evolutionary stable distributions of lekking ruffs (*Philomachus pugnax*).

Sutherland, W. J., and Parker, G. A. (1985). Distribution of unequal competitors. In R. M. Sibly and R. H. Smith, eds., *Behavioural Ecology: Ecological Consequences of Adaptive Behaviour*, pp. 255–274. Oxford: Blackwell.

Sutherland, W. J., and Parker, G. A. (1992). The relationship between continuous input and interference models of ideal free distributions of unequal competitors. *Anim. Behav., 44*, 345–356.

Svensson, B. G., and Petersson, E. (1987). Sex-role reversed courtship behaviour, sexual dimorphism and nuptial gifts in the dance fly *Empis borealis* (L.). *Ann. Zool. Fenn., 24*, 323–334.

Svensson, B. G., and Petersson, E. (1992). Why insects swarm: Testing the models for lek mating systems on swarming *Empis borealis* females. *Behav. Ecol. Sociobiol., 31*, 253–261.

Taylor, P. D., and Williams, G. C. (1982). The lek paradox is not resolved. *Theor. Pop. Biol., 22*, 392–409.

Théry, M. (1992). The evolution of leks through female choice: Differential clustering and space utilization in six sympatric manakins. *Behav. Ecol. Sociobiol., 30*, 227–237.

Thompson, C. W. (1991). The sequence of molts and plumages in painted buntings and implications for theories of delayed plumage maturation. *Condor, 93*, 209–235.

Thompson, S. (1986). Male spawning success and female choice in the mottled triplefin, *Forsterygion varium* (Pisces: Tripterygiidae). *Anim. Behav., 34*, 580–589.

Thornhill, R. (1980). Sexual selection within swarms of the lovebug, *Plecia arctica* (Diptera: Bibionidae). *Anim. Behav., 28*, 405–412.

Thornhill, R., and Alcock, J. (1983). *The Evolution of Insect Mating Systems*. Cambridge, Mass.: Harvard University Press.

Trail, P. W. (1985a). Territoriality and dominance in the lek-breeding Guianan cock-of-the-rock. *Natl. Geogr. Res., 1*, 112–123.

Trail, P. W. (1985b). Courtship disruption modifies mate choice in a lek-breeding bird. *Science, 227*, 778–780.

Trail, P. W. (1987). Predation and anti-predator behavior at Guianan cock-of-the-rock leks. *Auk, 104*, 496–507.

Trail, P. W. (1990). Why should lek-breeders be monomorphic? *Evolution, 44*, 1837–1852.

Trail, P. W., and Adams, E. S. (1989). Active mate choice at cock-of-the-rock leks: Tactics of sampling and comparison. *Behav. Ecol. Sociobiol., 25*, 283–292.

Trail, P. W., and Koutnik, D. L. (1986). Courtship disruption at the lek in the Guianan cock-of-the-rock. *Ethology, 73*, 197–218.

Treherne, J. E., and Foster, W. A. (1981). The effects of group size on predator avoidance in a marine insect. *Anim. Behav., 28*, 1119–1122.

Trillmich, F., and Trillmich, K.G.K. (1984). The mating systems of pinnipeds and marine iguanas: Convergent evolution of polygyny. *Biol. J. Linn. Soc., 21*, 209–216.

Trivers, R. L. (1972). Parental investment and sexual selection. In B. Campbell, ed., *Sexual Selection and the Descent of Man*, pp. 136–179. Chicago: Aldine.

Tuck, L. M. (1972). *The Snipes: A Study of the Genus Capella*. Ottawa: Canadian Wildlife Monographs.

Tuttle, M. D., Taft, L. K., and Ryan, M. J. (1982). Evasive behaviour of a frog in response to bat predation. *Anim. Behav., 30*, 393–397.

van Rhijn, J. G. (1973). Behavioural dimorphism in the ruff *Philomachus pugnax* (L.). *Behaviour, 47*, 153–229.

van Rhijn, J. G. (1983). On the maintenance and origin of alternative strategies in the ruff *Philomachus pugnax*. *Ibis, 125*, 482–498.

van Rhijn, J. G. (1991). *The Ruff*. London: T. and A. D. Poyser.

van Someren, V. D. (1947). The dancing display and courtship of Jackson's whydah *(Coliuspasser jacksoni* Sharpe). *E. Afr. Nat. Hist. Soc. J., 18*, 131–141.

Vehrencamp, S. L., and Bradbury, J. W. (1984). Mating systems and ecology. In J. R. Krebs and N. B. Davies, eds., *Behavioural Ecology: An Evolutionary Approach*, pp. 251–278. Oxford: Blackwell.

Vehrencamp, S. L., Bradbury, J. W., and Gibson, R. M. (1989). The energetic costs of display in male sage grouse. *Anim. Behav. 38*, 885–896.

Vellenga, R. E. (1970). Behaviour of the male satin bowerbird at the bower. *Austr. Bird Bander, 8*, 3–11.

Wade, M. J., and Pruett-Jones, S. G. (1990). Female copying increases the variance in male mating success. *Proc. Natl. Acad. Sci. USA, 87*, 5749–5753.

Wagner, R. H. (1992). Extra-pair copulations in a lek: The secondary mating system of monogamous razorbills. *Behav. Ecol. Sociobiol., 31*, 63–71.

Wagner, R. H. 1993. The pursuit of extra-pair copulations by female birds: A new hypothesis of colony formation. *J. theor. Biol., 163*, 33–346.

Walker, T. J. (1983). Mating modes and female choice in short-tailed crickets *(Anurogryllus arboreus)*. In D. T. Gwynne and G. K. Morris, eds., *Orthopteran Mating Systems: Sexual Competition in a Diverse Group of Insects*, pp. 240–267. Boulder: Westview Press.

Wallace, A. R. (1889). *Darwinism*. London: Macmillan.

Waltz, E. C. (1982). Alternative mating tactics and the law of diminishing returns: The satellite threshold model. *Behav. Ecol. Sociobiol., 10*, 75–83.

Warner, R. R. (1987). Female choice of sites versus mates in a coral reef fish, *Thalasoma bifasciatum*. *Anim. Behav., 35*, 1470–1478.

Warner, R. R. (1988). Traditionality of mating-site preferences in a coral reef fish. *Nature (London), 335*, 719–721.

Warner, R. R. (1990). Male versus female influences on mating-site determination in a coral reef fish. *Anim. Behav., 39*, 540–548.

Warner, R. R., and Schultz, E. T. (1992). Sexual selection and male charecteristics in the blue-headed wrasse, *Thallasoma bifasciatum*: Mating site acquisition, mating site defense, and female choice. *Evolution, 46*, 1421–1442.

Watts, C. R. (1968). Rio Grande turkeys in the mating season. *Trans. N. Am. Wildl. Conf., 23*, 205–210.

Wedell, N., and Arak, A. (1989). The wartbiter spermatophore and its effect on female reproductive output (Orthoptera: Tettigonidae, *Decticus verrucivorus*). *Behav Ecol. Sociobiol., 24*, 117–125.

Wegge, P., and Rolstad, J. (1986). Size and spacing of capercaillie leks in relation to social behaviour and habitat. *Behav. Ecol. Sociobiol., 19*, 401–408.

Wells, K. D. (1977). The social behaviour of anuran amphibians. *Anim. Behav., 25*, 666–693.

Wells, K. D., and Taigen, T. L. (1986). The effects of social interactions on calling energetics in the grey treefrog (*Hyla versicolor*). *Behav. Ecol. Sociobiol., 19*, 9–18.

Westcott, D. 1993. Habitat characteristics of lek sites and their availability for the ochre-bellied flycatcher, *Mionectes oleagineus. Biotropica, 25*, 444–451.

Westcott, D. (In prep.). Leks of leks suggest a role for hotspots in lek evolution.

West-Eberhard, M. J. (1979). Sexual selection, social competition and evolution. *Proc. Am. Phil. Soc., 123*, 222–234.

West-Eberhard, M. J. (1983). Sexual selection, social competition, and speciation. *Q. Rev. Biol., 58*, 155–183.

Whitney, C. L., and Krebs, J. R. (1975). Mate selection in Pacific tree frogs. *Nature (London), 255*, 325–326.

• Wicklund, C. G., and Andersson, M. (1980). Nest predation selects for colonial breeding among fieldfares *Turdus pilaris. Ibis, 122*, 363–366.

Wickman, P.-O. (1985). Territorial defence and mating success in males of the small heath butterfly, *Coenonympha pamphilus* L. (Lepidoptera: Satyridae). *Anim. Behav., 33*, 1162–1168.

Wickman, P.-O. (1986). Courtship solicitation by females of the small heath butterfly, *Coenonympha pamphilus* (L.) (Lepidoptera: Satyridae) and their behaviour in relation to male territories before and after copulation. *Anim. Behav., 34*, 153–157.

Wickman, P.-O., Garcia-Barros, E., and Rappe-George, C. In press. The location of landmark leks in the small heath butterfly, *Coenonympha pamphilus*: evidence against the hotspot model. *Behav. Ecol.*

Wiley, E. O. 1981. *Phylogenetics*. New York: Wiley & Sons.

Wiley, R. H. (1971). Song groups in a singing assembly of little hermits. *Condor, 73*, 28–35.

Wiley, R. H. (1973). Territoriality and non-random mating in sage grouse, *Centrocercus urophasianus. Anim. Behav. Monogr., 6*, 87–169.

Wiley, R. H. (1974). Evolution of social organization and life-history patterns among grouse. *Q. Rev. Biol., 49*, 201–227.

Wiley, R. H. (1991). Lekking in birds and mammals: Behavioral and evolutionary issues. *Adv. Study Behav., 20*, 201–291.

Willebrandt, T. (1988). Demography and ecology of a black grouse (*Lyrurus tetrix L.*) population. Ph.D. thesis, Uppsala University.

Williams, G. C. (1957). Pleiotropy, natural selection and the evolution of senescence. *Evolution, 11*, 398–411.

Williams, G. C. (1992). *Natural Selection: Domains, Levels and Challenges*. Oxford: Oxford University Press.

Williams, J. B., and Prints, A. (1986). Energetics of growth in savannah sparrows:

A comparison of doubly-labeled water and labortatory estimates. *Condor, 88,* 74–83.

Willis, E. O. (1966). Notes on a display and nest of the club-winged manakin. *Auk, 83,* 475–476.

Willis, E. O., and Eisenmann, E. (1979). A revised list of birds of Barro Colorado island, Panama. *Smiths. Contr. Zool., 291,* 1–31.

Willis, M. A., and Birch, M. C. (1982). Male lek formation and female calling in a population of the arctiid moth *Estigmene acrea. Science, 218,* 168–170.

Wittenberger, J. F. (1978). The evolution of mating systems in grouse. *Condor, 80,* 126–137.

Wittenberger, J. F. (1979). The evolution of mating systems in birds and mammals. In P. Marler, and J. Vandenbergh, eds., *Handbook of Behavioral Neurobiology,* vol. 3. New York: Plenum Press.

Wittenberger, J. F. (1981). *Animal Social Behavior.* Boston: Duxbury Press.

Wolf, L. L. (1970). The influence of seasonal flowering on the biology of some tropical hummingbirds. *Condor, 72,* 1–14.

Wrangham, R. W. (1980). Female choice of least costly males: A possible factor in the evolution of leks. *Z. Tierpsychol., 54,* 357–367.

Zahavi, A. (1975). Mate selection-a selection for a handicap. *J. theor. Biol., 53,* 205–214.

Zahavi, A. (1977). The costs of honesty (further remarks on the handicap principle). *J. theor. Biol., 67,* 603–605.

Zahavi, A. (1991). On the definition of sexual selection, Fisher's model and the evolution of waste and of signals in general. *Anim. Behav., 42,* 501–503.

Zucker, N. (1981). The role of hood-building in defining territories and limiting combat in fiddler crabs. *Anim. Behav., 29,* 387–395.

Zucker, N. (1983). Courtship variation in the neo-tropical fiddler crab *Uca deichmanni*: Another example of female incitation to male competition? *Mar. Behav. Physiol., 10,* 55–79.

Zuk, M., Johnson, K., Thornhill, R., and Ligon, J. D. (1990a). Mechanisms of female choice in red jungle fowl. *Evolution, 44,* 477–485.

Zuk, M., Johnson, K., Thornhill, R., and Ligon, J. D. (1990b). Parasites and male ornaments in free-ranging and captive red jungle fowl. *Behaviour, 114,* 232–248.

Zuk, M., Thornhill, R., Ligon, J. D., and Johnson, K. (1990c). Parasites and mate choice in red jungle fowl. *Am. Zool., 30,* 235–244.

Zwickel, F. C., and Bendell, J. F. (1972). Blue grouse, habitat, and populations. *Proc. 15th Int. Ornith. Congr.,* 150–169.

Author Index

Klump, G. M., 83
Knapton, R. W., 43, 66, 87
Kodrick-Brown, A., 44, 100
Köhler, D., 25
Koivisto, I., 69, 84, 124, 152, 157
Korpimäki, E., 98, 104
Koutnik, D. L., 71, 193
Krebs, J. R., 6, 31, 162
Kruijt, J. P., 54, 57, 58, 73, 74, 110, 117, 118, 133, 166, 197

Lack, D., 6, 137, 152
Lancingerova, J., 40
Landau, H. G., 69
Lande, R., 34, 95, 97, 101, 108, 142
Landel, H., 69, 80
Langbein, J., ix, 23, 39, 40, 155, 176, 177
Lank, D. B., 166, 167, 178, 181, 197, 205
Larsson, F., 158
Lazarus, J., 153, 154, 164
Lazenby-Cohen, K. A., 45
Leader-Williams, N., 45
LeBouef, B. J., 52, 53, 81
Le Croy, M., 80
Lederhouse, R. C., 5, 26, 43
Le Menec, M., 105
Lemeshow, S., 128
Lemnell, P. A., 8, 46, 147
Lewis, R. A., 10, 45, 145
L'Hospitalier, R., 105
Ligon, J. D., 76, 104
Lill, A., 35, 47, 53, 70, 73, 111, 118, 159
Lindell, L.-E., 58, 59, 68, 75, 81, 82, 172, 182
Lloyd, J. E., 5, 26, 34, 40
Lloyd, L., ix,
Løfaldli, L., 37, 57, 137
Loiselle, P. V., 28
Losey, G. S., Jr., 116
Lott, D. F., 175, 177
Louda, S. M., 11, 29, 30, 44, 59, 68
Lowe, R. H. *See* Lowe-McConnel
Lowe-McConnel, R. H., 29, 44
Loye, J. E., 76
Lucas, H. L., Jr., 184, 185
Ludwig, A., 69
Lumsden, H. G., 69, 117
Lund, M. K., 76, 77
Lundberg, A., xii, 3, 52, 53, 54, 57, 58, 59, 68, 69, 71, 73, 75, 76, 77, 78, 85, 89, 103, 104, 106, 112, 114, 115, 116, 117, 118,122, 124, 131, 132, 133, 136, 147,

152, 164, 165, 168, 169, 172, 181, 182, 192, 197
Lyon, B. E., 147, 178

MacColl, A., 5, 78, 108, 169, 170, 193, 197
MacNally, R. C., 27, 167
Majima, T., 43
Malavasi, A., 43
Manly, B.F.J., 77, 82
Manning, J. T., 75
Mari, F., 58, 59, 63, 64, 69, 70, 155, 176, 177
Marius-Jestin, V., 105
Markus, M. B., 104
Maynard-Smith, J., 94, 101, 102, 178, 183
McAndrew, F. T., 43
McCarthy, C. E., 153
McComb, K., 111
McCullagh, P., 128
McDonald, D. B., 53, 59, 68, 80, 88
McKaye, K. M., 11
McKaye, K. R., 14, 29, 30, 44, 59, 68
McKitrick, M. C., 35
Mclachlan, A. J., 153, 154, 164
McLennan, D. A., 148
McMinn, H., 76
Medawar, P. B., 73
Meixner, A. J., 27
Merton, D. V., 46
Michod, R. E., 102
Mika, J., 40
Milinski, M., 76, 184
Mjelstad, H., 105, 117
Moisan, J. C., 105
Møller, A. P., 6, 76, 93, 103, 107, 117, 119, 120, 136
Montgomerie, R. D., 8, 17, 147, 160, 164, 168, 180, 181, 182, 192, 197
Moore, A., 48
Morgante, J. S., 43
Morris, G. K., 12, 26, 27
Morris, R. V., 46
Moyer, J. T., 44
Murray, B. G., Jr., 137
Myers, J. P., 46

Nee, S., 98, 108, 204
Neems, R. M., 153, 154, 164
Nefdt, R.J.C., 58, 65, 84, 111, 160
Nelder, J. A., 128
Nichols, R. A., 97
Nilsson, S., 34
Nishida, T., 164

Subject and Species Index

Jacob Höglund is Associate Professor in the Department of Zoology at Uppsala University in Sweden. Rauno V. Alatalo is Professor in the Department of Biology at the University of Jyväskylä in Finland.